A Mathematical Tour

A Mathematical Tour introduces readers to a selection of mathematical topics chosen for their centrality, importance, historical significance, and intrinsic appeal and beauty. The book is written to be accessible and interesting to readers with a good grounding in high school level mathematics and a keen sense of intellectual curiosity. Each chapter includes a short history of the topic, statements and discussion of important results, illustrations, user-friendly exercises, and suggestions for further reading. This book is intended to be read for pleasure but could also be used for a Topics course in Mathematics or as a supplementary text in a History of Mathematics course.

Features

- contains a selection of accessible mathematical topics
- exercises that elucidate, and sometimes enlarge on, the topics
- suitable for readers with knowledge of high school mathematics

Denis Bell is a professor of mathematics at the University of North Florida. He was born and raised in London, England and studied at the Universities of Manchester and Warwick. His scholarship has been recognized by national funding and a Research Professorship at *MSRI*, Berkeley. He is also a writer of short and flash fiction, with numerous pieces in print in literary journals in the United States and elsewhere.

Chris Bernhardt is a professor emeritus of mathematics at Fairfield University, where he taught for 30 years. He was born and raised in England and studied at the University of Warwick. His books include *Turing's Vision: The Birth of Computer Science, Quantum Computing for Everyone* and *Beautiful Math.*

A Mathematical Tour

Denis Bell and Chris Bernhardt

CRC Press
Taylor & Francis Group
Boca Raton London New York

CRC Press is an imprint of the
Taylor & Francis Group, an **informa** business
AN A K PETERS BOOK

First edition published 2025
by CRC Press
2385 NW Executive Center Drive, Suite 320, Boca Raton FL 33431

and by CRC Press
4 Park Square, Milton Park, Abingdon, Oxon, OX14 4RN

CRC Press is an imprint of Taylor & Francis Group, LLC

ISBN: 978-1-032-74789-7 (hbk)
ISBN: 978-1-032-74738-5 (pbk)
ISBN: 978-1-003-47091-5 (ebk)

DOI: 10.1201/9781003470915

Typeset in Latin Modern font
by KnowledgeWorks Global Ltd.

Publisher's note: This book has been prepared from camera-ready copy provided by the authors.

Contents

Preface

The famous twentieth century mathematician Paul Erdős spoke of "The Book," a tome existing in heaven where God collects the most elegant mathematical theorems. The theme of beauty in mathematics is expounded at some length in the prominent Oxford mathematician G. H. Hardy's autobiography of 1940, "A Mathematician's Apology." Hardy contended that only beautiful mathematics has a right to exist and, moreover, the mathematics that falls into this category is almost exclusively of the "pure" variety, existing for its own sake and devoid of any real-world applications.[1] Hardy characterizes what he considers the best mathematical proofs as something akin to great music and poetry, exhibiting a seemingly paradoxical combination of inevitability and surprise. He cites as examples Euclid's proof of the infinitude of the prime numbers and the Pythaoreans' proof of the irrationality of $\sqrt{2}$.

Hardy notwithstanding, mathematics is valued today both for its esoteric qualities and as a tool that has played a fundamental role in creating the modern world. Viewed in this way, the discipline appears as one of the most magnificent creations of the human mind. But mathematics also has a human dimension. The men and women who created this vast body of work were just that, men and women. They had wives, husbands, children, or in some cases, they didn't. They were gamblers, philanderers, judges, clerics, soldiers, poets, writers, and rakes. They suffered, they triumphed, they starved, they courted the favor of nobles, they lorded it over each other at times, some were even killed. It is the purpose of this book is to present mathematics in both its humanistic and scientific aspects, although being mathematicians rather than historians, our focus is largely on the latter.

So, welcome to our tour! There are indeed beautiful vistas. If the terrain gets a little rocky in places, don't worry, just stay close and let us be your guide. The landscape is divided into subject areas. The first two chapters are on Geometry and Number Theory. Chapter 3, on Medieval and Renaissance Mathematics, is transitionary in nature, bridging the gap between "old mathematics" and the invention of the subject's greatest tool, calculus. Chapter 4

[1] In fairness, it should be noted that as a lifelong pacifist, Hardy's jaundiced view of the applied side of the subject was no doubt partly due to abhorrence at the use to which mathematics was put in the service of war. It is unfortunate that Hardy never learned of the work of Alan Turing and others at Bletchley Park during World War II who, by applying mathematical techniques to codebreaking, probably shortened the war by 2 years and thereby saved millions of lives.

concerns Algebra, told through the history of polynomial equations. Chapter 5 provides an introduction to Calculus proper, with discussions of the main ideas and theorems. Chapter 6 is an account of Complex Variables, the calculus of functions defined using complex numbers. Chapter 7 addresses Graph Theory, a subject that originated with Euler's solution to the bridges of Königsberg problem, and which has grown into a major branch of mathematics, with numerous applications in areas ranging from economics to computer networks. Chapter 8 is concerned with Probability and traces the development of the subject from its ancient roots in gambling to its role as the foundation of mathematical statistics. Finally, Chapter 9 deals with the subject of the infinite, mathematical logic, and elements of computer science, from the theory of countability originated by Cantor, through Gödel's revolutionary work on undecidable propositions, to Turing's work on computing machines and computable numbers.

The background required to read this book is relatively modest; high school algebra and trigonometry should suffice for most of it. We believe the book will serve a variety of purposes: as a text in a Topics in Mathematics course, as a supplementary text in History of Mathematics, as an introduction for graduating high school students to the type of mathematics they are likely to encounter in college, or simply as a book of interest to those who know some mathematics and would like to learn more. In any event, we can only hope that the book will prove as much fun to read as it was to write. Please let us know your thoughts on this matter, if you are so inclined.

Denis Bell (dbell@unf.edu)
Chris Bernhardt (chris@chrisbernhardt.info)

Acknowledgments

The authors wish to thank the anonymous reviewers for their careful reading of the manuscript, pointing out our many errors and providing insightful suggestions. Likewise, Peter Schumer, Michelle Dedeo, Cynthia Bell, Madison Bell, Mitchel Burchfield, David Burchfield, and Brian Sunguza. Ian Stewart, Jeanne-Pierre Tignol, and David Burton, whose books we drew on, together with Wikipedia. The staff at Taylor & Francis, Acquiring Editor, Callum Fraser, and Editorial Assistant, Mansi Kabra, for their kindly assistance in bringing this project to fruition.

Introduction: A Roadmap

The book is organized along the lines of topics in mathematics. For this reason, the presentation is somewhat ahistorical, although the arrangement of material within each chapter is chronological (as far as proved possible while maintaining a coherent narrative structure).

The first two chapters deal with the most ancient of mathematical disciplines, geometry and number theory. The greatest achievement of the Greeks, Euclid and his mathematical descendants, was perhaps to establish what became the paradigm for mathematical thought over the centuries: that mathematical truths need to be proved by logical arguments founded upon a set of basic premises. Euclid's style of mathematics is laid out in that great foundational work *The Elements* in the form of a set of *axioms* (the premises), *theorems* (the truths), and *proofs* (the logical arguments). On the basis that one could do no better than emulate the master, we chose to follow Euclid's style in Chapters 1 and 2. Here the reader will find a series of mathematical constructions, theorems, and proofs. While the proofs are certainly of interest in their own right and would repay careful reading, they could be skipped over in a casual reading of the book.

The basic objects treated in Chapters 1 and 2, namely shapes and numbers, occur naturally in the world, so to speak, and therefore required no formulation. In order to extend Euclid's mode of thought to more abstract areas of mathematics such as algebra, it was necessary first to devise a language with which to express the ideas, then an efficient notational system to facilitate their development. This came about in the Renaissance period with the introduction of symbolic algebra (equations with definite but unspecified coefficients, an unknown quantity to be solved for, etc.). This, together with the invention of analytic geometry (representing curves by equations), allowed algebra to flourish as an independent discipline and paved the way for the invention of calculus in the seventeenth century. These developments, and many other important contributions of this era, are presented in Chapter 3.

Chapter 4 is devoted to algebra, told primarily through the history of polynomial equations, starting with quadratic equations, through cubic and higher degree equations, and culminating in the more abstract developments of the nineteenth and early twentieth centuries. Along the way, we survey complex numbers and discuss the Fundamental Theorem of Algebra and Gauss' construction of the 17-sided polygon.

Chapter 5 presents the elements of Calculus, starting with a brief survey of the history of the subject and its founders, Newton and Leibniz. Subsequent sections expound upon on the two branches of the subject, Differentiation and Integration and their melding in the Fundamental Theorem of Calculus. One of the most spectacular achievements of mathematics in the seventeenth and eighteenth centuries was the representation of functions such as sine, cosine, exponential, and logarithm, as infinite series, i.e., polynomials with an infinite number of terms. A discussion of infinite series, as developed by Taylor, Maclaurin, Lagrange, and others, is presented in the final section of Chapter 5.

By this point in the book we abandon the somewhat spartan approach assumed in the early chapters in favor of a more informal style, generally presenting theorems without proofs. This is so as to allow readers to become acquainted with the elegance of the mathematics without getting bogged down in technical details. Needless to say, this mode of presentation is no substitute for a rigorous study of the topics. Suggestions for further reading in this direction are provided at the end of every chapter.

Chapter 6 is an introduction to the subject of Complex Variables, the calculus of functions of complex (as opposed to real) numbers. While incorporating much of the theory in real variable calculus, the subject has a flavor all its own and offers up a wealth of new and important results, including some that throw new light on real-valued functions. An example is a powerful new method to evaluate integrals and infinite sums.

Chapter 7 introduces the reader to the subject of Graph Theory. We first discuss the origins of the subject in the bridges of Königsberg problem. The next sections introduce the main object of study, planar graphs, discuss their properties, and introduce a number known as the *Euler characteristic* which captures an essential feature of the graphs. The chapter concludes with an application of graph theory in three dimensions and a theorem of Gauss and Bonnet which establishes a remarkable connection between the Euler characteristic of a surface and the geometry of the surface.

The final two chapters of the book reflect the mathematical interests of the two authors. The subject of Chapter 8 is Probability. The chapter opens with a discussion of the history of the subject in games of chance, then moves on to describe various methods to calculate probabilities and applies these in some classic situations, demonstrating, for example, that you are unlikely to get rich by playing the slots in Vegas. We point to the role of probability as the foundation for mathematical statistics, discuss laws of large numbers, and indicate how probabilistic methods can be used to prove a rather surprising fact about numbers.

Chapter 9 deals with the mathematical theory of the infinite and related matters in the theory of computation. Some highlights of this chapter are an argument of Cantor showing how to distinguish different types of infinities, Gödel's incompleteness theorem that any consistent mathematical system necessarily contains propositions which can neither be proved nor disproved, and Turing's work which became the theoretical basis for computer science and indeed for the device that I am using right now! The book concludes with the discussion of a notorious unsolved problem in the theory of computation, "P versus NP."

Now for the promised roadmap. We expect that a large portion of this book will be accessible to readers with a solid grounding in high school geometry and algebra. We suggest in this regard:

Chapter 1

Chapter 2 (except Section 2.12)

Chapter 3 (except Section 3.6)

Chapter 4 (Sections 4.1–4.6)

Chapter 5 (except Section 5.6)

Chapter 7 (except Section 7.7)

Chapter 8 (Sections 8.1–8.3)

Chapter 9

The other sections are a little more technically demanding and need more background, in particular a first course in calculus, would be useful. Relevant material is provided in Chapter 5.

Readers are invited to "cherry-pick" from the book. For example, those with a geometric or visual bent might enjoy Chapters 1 and 7; numerologists and algebraists, Chapters 2 and 4; for the analytically minded, Chapters 5 and 6; Chapter 8, for readers who plan to join the professional poker circuit; Chapter 9, for those interested in modern developments ...

With the exception of Chapters 5 and 6, which obviously are interlinked, the chapters are largely independent. There are exercises sprinkled throughout to aid in understanding, with solutions provided to the starred exercises. Also, as remarked above, suggestions for further reading. Happy touring!

CHAPTER 1

Geometry

The presentation in Chapters 1 and 2 is very much in the style of Euclid's *Elements*, as indicated in the Forward. Our intention here is to convey to the reader a sense of the style of mathematics of this era, which proved to be so influential in shaping the future development of the subject.

Chapter 1 begins with a discussion of probably the best known result in all of mathematics, the Pythagorean theorem. We present two simple proofs of the theorem, an algebraic proof and a geometric proof.

The *Elements* is based on five initial assumptions, called axioms. The most challenging of the axioms is the so-called *Parallel Postulate*, which asserts that a unique straight line can be drawn through any given point parallel to a given line. Two equivalent versions of the postulate are presented along with some of its consequences, including the proof that the angles in a triangle add up to two right angles (180°).

Euclid's axioms provide for the construction, using straightedge and compass, of the familiar geometric figures: circles, triangles, and rectangles. This topic is discussed in Section 1.3, along with other "Euclidean constructions," and the algebra underlying them. We describe the construction by straightedge and compass of the *golden ratio* and the regular pentagon. The subject of Euclidean constructions is briefly taken up again in Chapter 4, with Gauss' construction of the regular *heptadecagon* (17-sided polygon).

Starting in the eighteenth century, a succession of mathematicians sought to prove the Parallel Postulate from Euclid's other axioms. They were led to consider how geometry would look if the Parallel Postulate were dropped from the set of axioms. This resulted in the creation by Riemann, Bolyai, and Lobachevsky, of so-called non-Euclidean geometries, a subject which was later to find application in Einstein's General Theory of Relativity. We discuss these developments.

Section 1.4 is concerned with the contributions of Archimedes, widely considered to be the greatest mathematician of the ancient world. Included are some of Archimedes' achievements in the realm of engineering, his method of

DOI: 10.1201/9781003470915-1

approximating π by inscribed and circumscribed polygons, and his determination of areas and volumes by a process very similar to what we now know as calculus. The chapter concludes with a classification of the so-called Platonic solids.

1.1 THE PYTHAGOREAN THEOREM

There are few theorems that the average person on the street knows, but most know the Pythagorean theorem. Of all the results in mathematics, It is probably the most well-known and the most proved. There are literally hundreds of proofs of this theorem (even Napoleon came up with one!). In this section, we give a couple of them and then discuss some reasons why this theorem is so important.

The theorem concerns right triangles. In figure 1.1, we depict our triangle with the lengths of the sides denoted by a, b, and c.

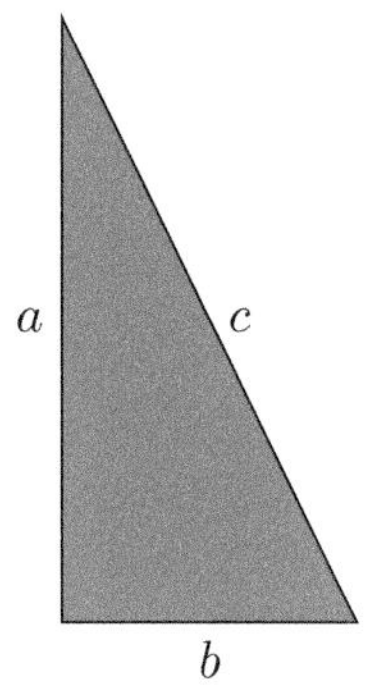

FIGURE 1.1 A right triangle

Theorem 1.1. *(Pythagorean) Given a right triangle, the area of the square on the hypotenuse equals the sum of the areas of the squares on the other two sides.*

Figure 1.2 shows the triangle with the squares in gray.

The history of this theorem dates well before the life of Pythagoras, who lived around 500 BCE. A clay tablet dating to 1800 BCE, known as Plimpton 322, contains a list of *Pythagorean triples*—numbers a, b, and c such that

$$a^2 + b^2 = c^2.$$

Chinese and Indian mathematicians independently discovered various special cases of the theorem, perhaps through floor tilings.

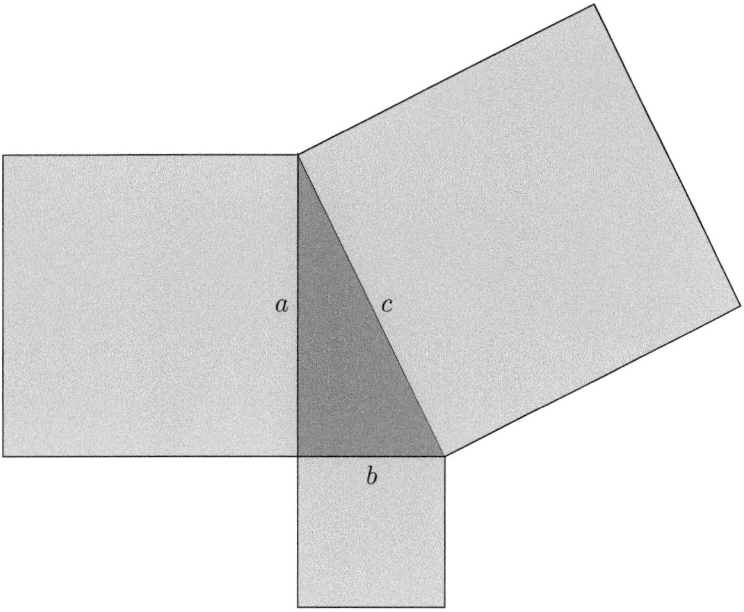

FIGURE 1.2 A right triangle with squares on the edges

We will give two proofs.

Algebraic Proof

Proof. We want to show that $a^2 + b^2 = c^2$. As a first step, we construct a square with sides of length $a + b$.

Inside this square we draw four of our right triangles as shown in figure 1.3.
The area of each of the triangles is $1/2(ab)$.
The area of the interior white square is c^2.
The area of the large square is $(a + b)^2$. This is equal to the sum of the areas of the white square and the four triangles. Writing this as an equation:

$$(a + b)^2 = c^2 + 4 \times 1/2(ab).$$

Expanding gives:

$$a^2 + 2ab + b^2 = c^2 + 2ab.$$

Subtracting $2ab$ from both sides gives the result we want to prove:

$$a^2 + b^2 = c^2.$$

□

This proof is both simple and elegant, but would seem quite mysterious to Euclid and Pythagoras. They did not have algebra. Fortunately, we can make a small change to our argument that avoids algebra and shows the three squares.

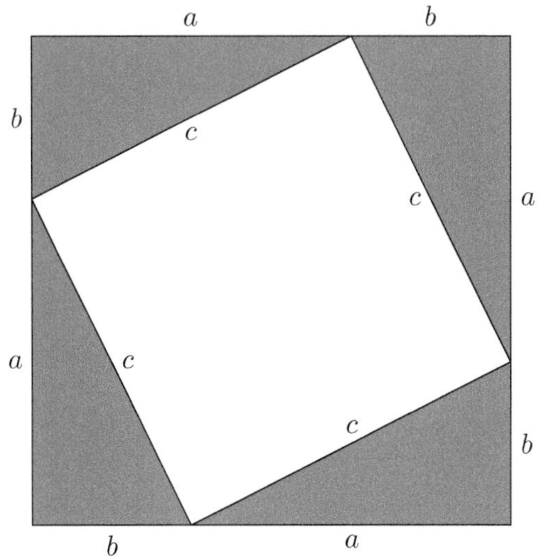

FIGURE 1.3 A square containing four copies of the right triangle

Geometric Proof

Proof. We start once more with our square with sides of length $a+b$ and the four triangles depicted in figure 1.3.

We can think of the four gray triangles as tiles lying on a white square with sides of length $a+b$. Now move the two triangles on the right side over to the triangles on the left so they form rectangles. Figure 1.4 shows the square with the new positions of the triangular tiles.

Finally, we observe that since the total gray area in both figures is the same, the total area in white must also be the same in both figures. The white area in figure 1.3 is the area of the square on the hypotenuse. The white area in figure 1.4 is the sum of the areas of the squares on the other two sides. □

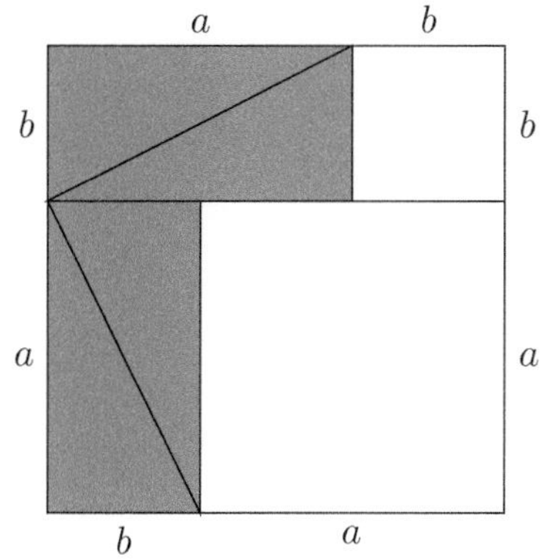

FIGURE 1.4 The large square with triangles moved to form rectangles

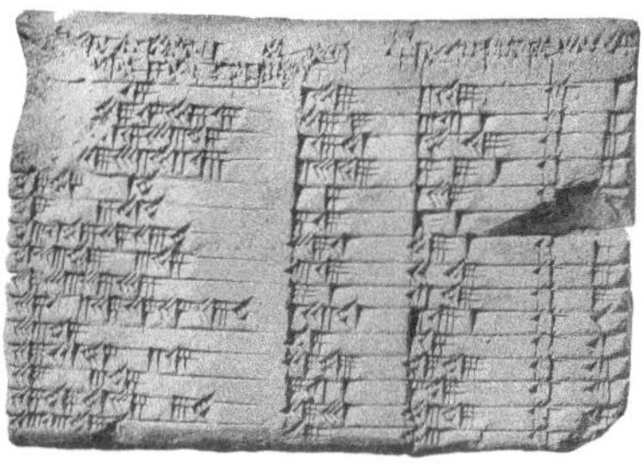

FIGURE 1.5 Plimpton 322

Distance between Points

The Pythagorean theorem tells us about areas of squares on the sides of right triangles, but the result gives a formula for the distance between two points in rectangular coordinates.

Fermat and Descartes independently invented analytic geometry in the first half of the seventeenth century, studying curves using rectangular coordinates. Coordinates in terms of longitude and latitude had been used much earlier, but in analytic geometry, we consider equations involving the coordinates. Once we have algebraic equations, we can use algebra to manipulate them.

Given two points, one with coordinates (x_0, y_0) and the other with (x_1, y_1). The distance between them is

$$\sqrt{(x_1 - x_0)^2 + (y_1 - y_0)^2}.$$

This follows immediately from the Pythagorean theorem, as illustrated in figure 1.6.

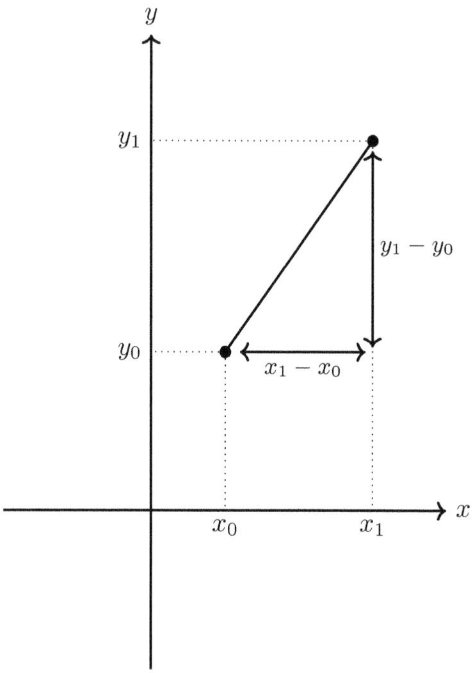

FIGURE 1.6 Distance between two points

Exercise 1.1.[⋆] *We must be careful to make sure that our diagrams don't mislead. In figure 1.6, the picture shows the case when $x_0 < x_1$ and $y_0 < y_1$. Show that the distance formula is still correct when $x_0 \geq x_1$ or $y_0 \geq y_1$.*

Exercise 1.2. *A circle of radius r centered at a point with coordinates (h, k) is defined to be the set of points whose distance to (h, k) is r. Deduce that the equation of the circle is*

$$(x - h)^2 + (y - k)^2 = r^2.$$

Exercise 1.3. *Consider the half-line through the origin that makes an angle θ with the positive x-axis. This line intersects the unit circle. The coordinates of this point are $(\cos(\theta), \sin(\theta))$. (This is illustrated in figure 1.7.) Prove the identity,*

$$\cos^2(\theta) + \sin^2(\theta) = 1.$$

In mathematics we often use Δ to denote the change in a quantity, so if x changes from x_0 to x_1, we write $\Delta x = (x_1 - x_0)$. Our formula for distance in the plane becomes:

$$s = \sqrt{(\Delta x)^2 + (\Delta y)^2}.$$

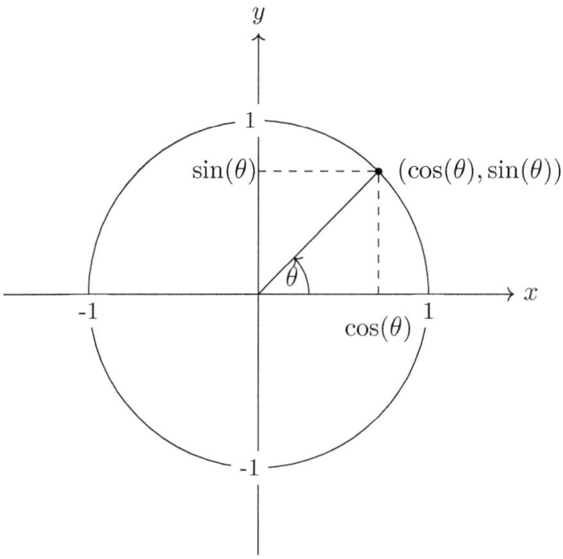

FIGURE 1.7 Unit circle with point $(\cos(\theta), \sin(\theta))$ shown

We can use the Pythagorean theorem repeatedly to find the distance formula in higher dimensions. In three dimensions, it is

$$s = \sqrt{(\Delta x)^2 + (\Delta y)^2 + (\Delta z)^2}.$$

We illustrate the proof in figure 1.8. First, the hypotenuse of the triangle with horizontal shading is calculated to be $\sqrt{(\Delta x)^2 + (\Delta y)^2}$. This is one side of the triangle with vertical shading, the other having length $|\Delta z|$. The hypotenuse of this second triangle gives the distance between the two points.

1.2 EUCLID AND THE PARALLEL POSTULATE

Euclid is, of course, known as the author of the *Elements*, that great edifice of Greek mathematics, and the founder of the axiomatic method. Euclid lived around 300 BCE, in the time between Plato and Archimedes. Though little is known about his life, we believe that Euclid studied at the Platonic Academy, later taught at the Musaeum, and spent much of his later career in Alexandria.

The *Elements* is built upon a foundation of five postulates, mathematical truths considered being so self-evident that they could be assumed at the outset. The first four postulates are as follows:

1. *A straight line can be drawn from any point to any other point.*
2. *A finite straight line can be produced continuously in a line.*

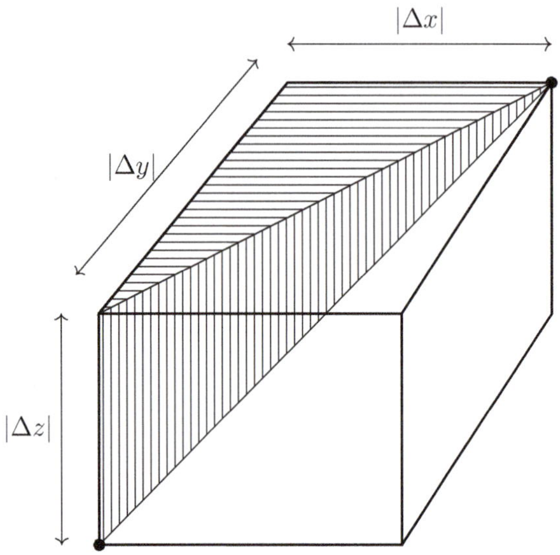

FIGURE 1.8 The distance between two points in three dimensions

FIGURE 1.9 Rendition of Euclid by Jusepe de Ribera, c. 1630–1635

3. *A circle may be described with any center and distance.*

4. *All right angles are equal to one another.*

The first three postulates allow the use of straightedge and compass in Euclidean constructions. The fifth of Euclid's postulates is the so-called Parallel Postulate:

5. *If a straight line falling on two straight lines makes the interior angles on the same side less than two straight right angles, then the two straight lines, if produced indefinitely meet on that side on which the angles are less than two right angles.* (see figure 1.10).

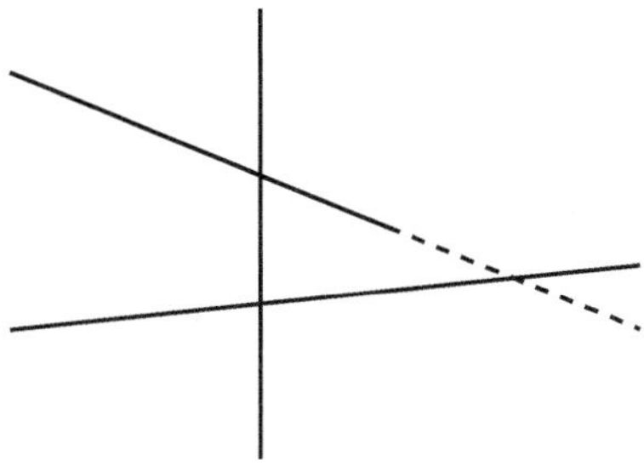

FIGURE 1.10 Euclid's Parallel Postulate

Euclid's statement of the parallel postulate can be replaced with Playfair's postulate: *Given any straight line l and a point p lying outside of l, there is a unique straight line passing through p and parallel to l* (see figure 1.11).

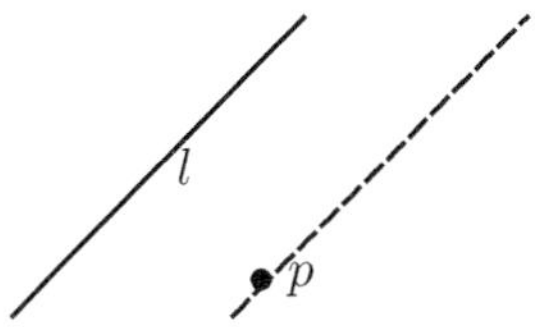

FIGURE 1.11 Playfair's postulate

The Parallel Postulate is the most remarkable of Euclid's postulates; for more than 2 millennia, it was thought that it could be proved from the other postulates in Euclidean geometry (and theorems derived from these), and thus did not require the status of a separate postulate. The thinking on this changed radically in the early 1800s with the creation of *non-Euclidean* geometries, i.e., internally consistent geometric systems, which *negate* the parallel postulate.

Giovanni Saccheri's work in 1733 started the change. Saccheri wanted to derive the parallel postulate from the other four postulates. Playfair's version of the fifth postulate states that given a line l and a point P not on l, there is a unique line through P parallel to l. Saccheri's approach was to negate this statement and try to derive a contradiction. There are two ways of negating Playfair's axiom, to stipulate:

1. There is no line through P that is parallel to l.
2. There is more than one line through P parallel to l.

Saccheri showed if the first case were true, then lines could not be continuously extended, contradicting Euclid's second postulate. Thus, he had ruled out the first case and he now sought to show that the second case also led to a contradiction. He was unable to do this. Gauss, Bolyai, and Lobachevsky further developed the second case and realized it gave a consistent new geometry that came to be known as *hyperbolic geometry.*

G. B. Riemann re-examined the first case of Saccheri's work. In doing so, he constructed a different type of non-Euclidean geometry, where the two-dimensional plane is replaced by the surface of a sphere. The analogue of line segments in this context are parts of great circles, i.e., circles formed by the intersection of the sphere with a plane passing through the center of the sphere. Portions of great circles are called *geodesics.* (In general, a geodesic on a surface is the curve of shortest length between its endpoints.) A triangle is a closed region of the sphere bounded by three geodesics, a quadrilateral by four, etc. There are no parallel lines in this theory because any two great circles intersect. Hence the first case of the negation of Playfair's postulate holds. Since the analogue of straight lines in this theory, i.e., great circles, are indeed finite in extent, Saccheri's contradiction ceases to apply, and the theory can be shown to be consistent with Euclid's first four postulates. Riemann's spherical geometry, in a slightly generalized guise, goes under the name of *elliptic geometry.*

Spherical geometry is an example of a geometry of constant curvature (the curvature at every point is equal to $1/R^2$, where R is the radius of the sphere. C. F. Gauss developed a comprehensive geometry of surfaces and Riemann extended Gauss's work to higher-dimensions. Riemannian geometry became a cornerstone of mathematics in the twentieth century and found its most spectacular application in Einstein's theory of general relativity.

We now prove the well-known fact in Euclidean Geometry that the angle sum of a triangle is 180°. The proof requires a preliminary result, which we state and prove first for the sake of completeness.

Theorem 1.2. *(Corresponding Angle Theorem) Suppose the lines labeled l and m in figure 1.12 are parallel. Then $\angle b = \angle c = \angle e$.*

Proof. We first prove the assertion $b = e$. We do this via an argument known as proof by contradiction (*reductio ad absurdum*), which goes as follows: assume that the theorem is false. Then seek to derive from this a contradiction, either to the premise of the theorem or to something else that has previously been proven. This contradiction proves that theorem must, in fact, be true.

Suppose, then, that $b \neq e$. Since $b + d = 180°$, this implies $d + e \neq 180°$. It follows from the Parallel Postulate that the lines l and m in figure 1.12 intersect, contrary to our assumption that they are parallel. This contradiction proves that $b = e$. To complete the proof, we observe that both $a + b = 180°$ and $a + c = 180°$, so $b = c$.

□

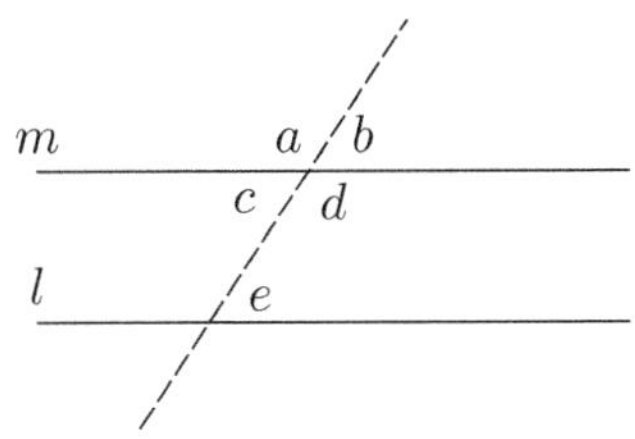

FIGURE 1.12 Supplementary and interior angles

Theorem 1.3. *(Angle sum in a triangle) The sum of the three angles in any triangle is* $180°$.

Proof. Given a triangle with vertices A, B, and C and angles a, b, and c, extend the line AC. Then draw a line through C parallel to AB. This is shown in figure 1.13. Notice that e and a are corresponding angles, as are b and d, so Theorem 1.2 tells us $d = b$ and $e = a$ Thus $a + b + c = c + d + e = 180°$. □

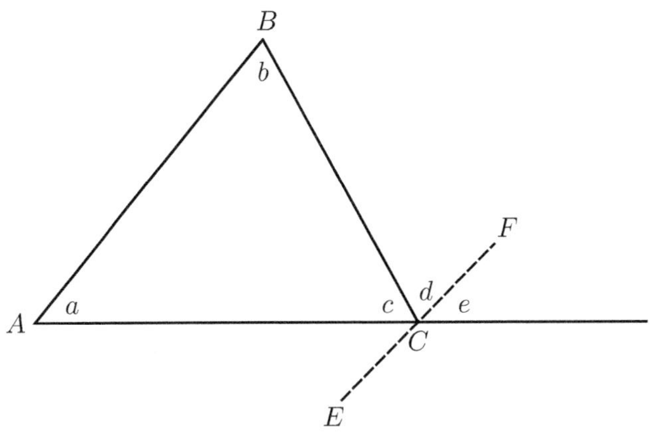

FIGURE 1.13 Proof of the angle sum theorem

Note that in the theorem's proof, we repeatedly use the Parallel Postulate. *The Angle Sum Theorem is not true in non-Euclidean geometries.* Behold the triangle ABC shown in figure 1.14, where the angle sum is 270°!

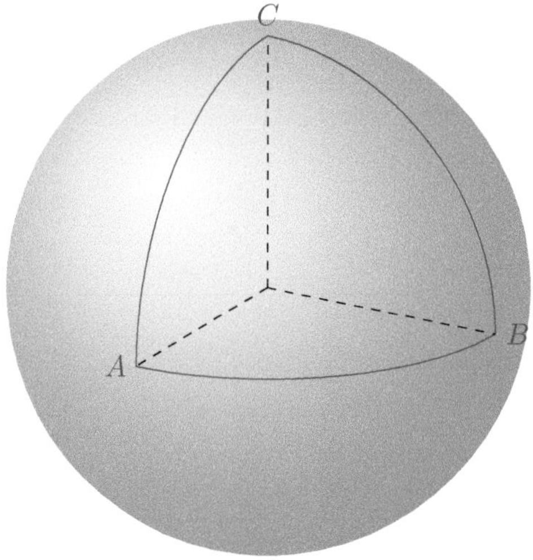

FIGURE 1.14 A triangle with angle sum 270°

1.3 EUCLIDEAN CONSTRUCTIONS

Euclid and his contemporaries' conception of number was intimately tied to their geometry; outside of the integers and rationals (fractions), numbers existed for them only in so far as they arose (as lengths) in geometric constructions using straightedge and compass, as allowed by Euclid's first three postulates. We call such numbers *constructible* and in this section describe a series of rules and constructions for their generation.

Theorem 1.4. *Suppose that a and b are constructible numbers. Then $a+b, a-b, ab$, and a/b are all constructible. Furthermore, $\sqrt{a}$ is constructible.*

Proof. The constructibility of $a+b$ and $a-b$ are clear. The others come from similar triangles, as we now explain. We start with an observation, the validity of which is evident from figure 1.15: *we can construct a right angle at any given point on a straight line.* This allows us to construct a right triangle having legs of prescribed length.

Figure 1.16 illustrates the construction of ab. By similar triangles, we have

$$\frac{1}{b} = \frac{a}{x},$$

hence $x = ab$.

The construction of a/b follows similar lines (no pun intended). □

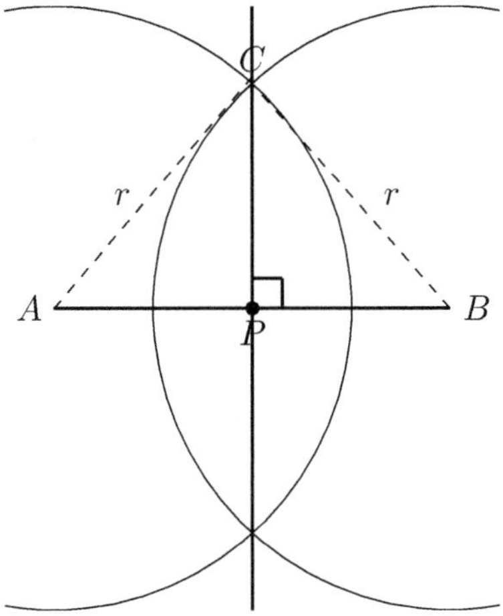

FIGURE 1.15 Construction of a perpendicular

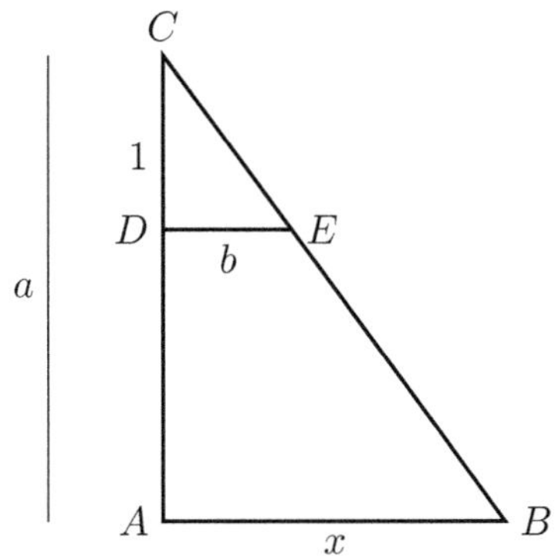

FIGURE 1.16 Construction of *ab*

The construction of $\sqrt{a}$ is illustrated in figure 1.17. Draw a circle with radius $a + 1$ and make the perpendicular at D as shown. Then the vertex angle at C is 90°.

Exercise 1.4.★ *Prove this using the angle sum of the two triangles formed by drawing the radius from the center of the semi-circle to C.*

Hence, $\triangle ACD$, $\triangle ABC$, and $\triangle BCD$ are similar. Identifying corresponding sides in $\triangle ACD$ and $\triangle BDC$, we have

$$\frac{x}{a} = \frac{1}{x},$$

thus $x = \sqrt{a}$.

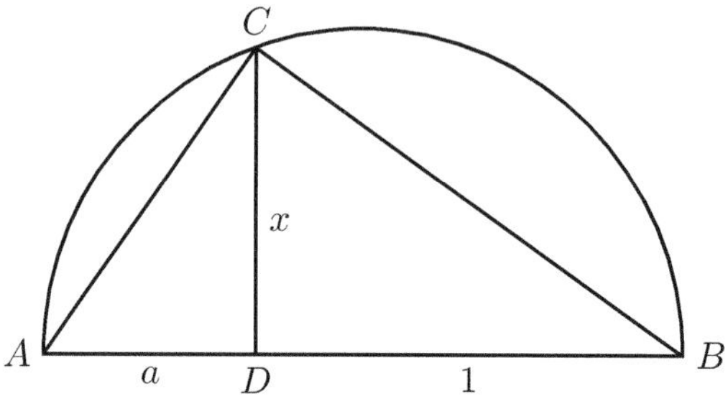

FIGURE 1.17 Construction of a square root

It follows from Theorem 1.4 that any number formed from the integers by performing successively any (finite) number of applications of the arithmetic operations and square roots is constructible. For example, the following number is constructible.

$$\sqrt{\sqrt{5}+\sqrt{20+\sqrt{\frac{\sqrt{10}-\sqrt{7}}{2}}}}$$

Geometric Solution to Quadratic Equations

Euclid devised a construction for finding the solutions to the quadratic equation[1]

$$x^2 + b^2 = ax. \tag{1.1}$$

Construct a line segment AB of length a, and the perpendicular PC with length b at the midpoint P (as indicated in figure 1.18). Draw a circle with center C and radius $a/2$. Let D be the intersection of this circle with AB (we are assuming here that $a > 2b \geq 0$).

Then the two roots of equation 1.1 are $|AD|$ and $|DB|$.

Exercise 1.5.* *Prove this algebraically.*

[1] Euclid, of course, did not state this problem using algebra. He stated it in terms of the areas of two squares and a rectangle.

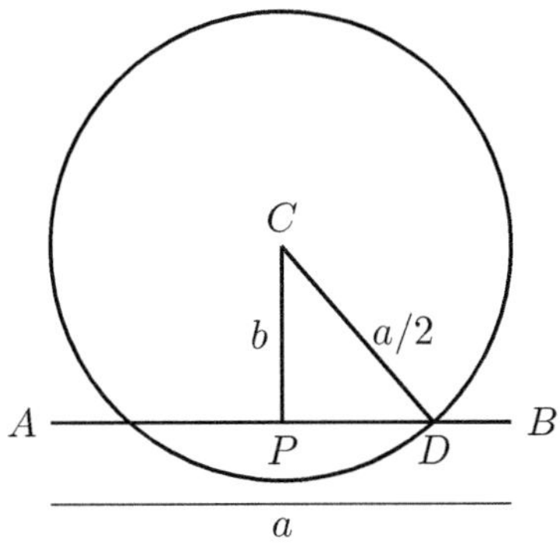

FIGURE 1.18 Geometric solution of the quadratic equation

Further Classical Constructions: Golden Ratio

The golden ratio arises from the classical problem: *divide a segment of unit length into two parts so that the ratio of the whole to the larger part is equal to the ratio of the larger part to the smaller.* Denote the lengths of the two parts by a and b with a the larger part, and the ratio by ϕ. Then

$$\phi = \frac{a+b}{a} = \frac{a}{b}.$$

This ratio is called the *golden ratio.*

One way to visualize this is with a rectangle with sides $a+b$ and a. Figure 1.19 depicts it. Removing a square of side a from one end of the rectangle, results in a smaller rectangle with sides a and b (this is the gray rectangle in the figure). Since,

$$\frac{a+b}{a} = \frac{a}{b},$$

the large rectangle and the smaller rectangle have sides in the same ratio.

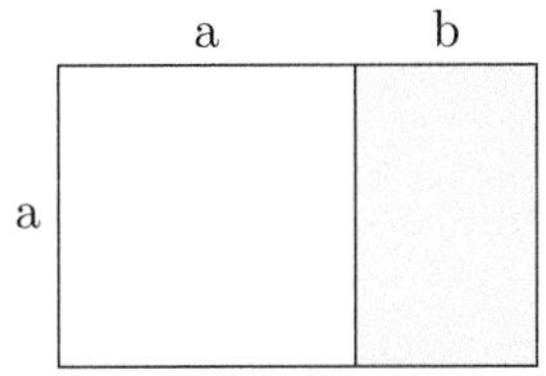

FIGURE 1.19 Golden rectangle

We call rectangles whose sides are in this ratio *golden.* Golden rectangles have the interesting property that when you delete a square containing the shorter edge, you end up with another golden rectangle. Many people consider golden rectangles to be the rectangles with the most aesthetically pleasing shape. A book published in 1509, *Divina proportione*, written by Luca Pacioli[2] and illustrated by Leonardo da Vinci became very influential in this regard and many works of the great Renaissance artists and designs of classical buildings such as the Parthenon have since been analyzed showing proportions in the golden ratio. The golden ratio is also closely related to a famous sequence of integers known as the *Fibonacci sequence*, which we will discuss later in Chapter 4.

We know ϕ satisfies

$$\phi = 1 + \frac{1}{\phi}.$$

Writing this as the quadratic equation $\phi^2 = 1 + \phi$ and solving gives

$$\phi = \frac{1 + \sqrt{5}}{2}.$$

According to the previous discussion, ϕ is constructible.

We now show how to construct a golden rectangle using straightedge and compass. Figure 1.20 illustrates it.

1. Draw a square. Label the vertices A, B, C, and D as in the figure.
2. Find the midpoint of edge AB and label it E.
3. Draw a circle centered at E that passes through C.
4. Extend line AB until it meets the circle at F.
5. Draw rectangle with sides AF and AD.

To see why this works, it is easiest if we think of the initial square as having sides of length 2. Then the radius of the circle is the hypotenuse of the triangle EBC and so equals

$$\sqrt{(1^2 + 2^2)} = \sqrt{5}.$$

This gives the length of AF as $1 + \sqrt{5}$ and AD as length 2.

[2]Pacioli translated an earlier manuscript by Piero della Francesca on the geometry of polyhedra and included this translation as part of his book. He did not credit della Francesca. This is one of the earliest known cases of plagiarism!

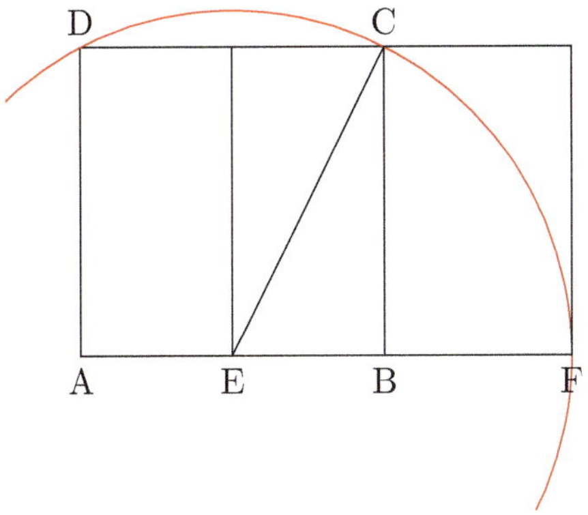

FIGURE 1.20 Construction of golden rectangle

Further Classical Constructions: Regular Polygons

A *regular polygon* is a polygon that has edges of the same length and all interior angles equal. The regular polygon with three edges is the equilateral triangle. A construction of this, given in *The Elements*, is obtained by choosing $r = |AB|$ in figure 1.15.

The regular polygon with four edges is the square. This is straightforward to construct using our construction for perpendiculars. See figure 1.15 once more.

Once you have a regular polygon with n sides, it is easy to construct one with $2n$ sides. This is done by drawing a circumscribing circle passing through all the vertices, then finding the intersections of the perpendicular bisectors of the edges with the circle. These intersections along with the vertices of the original polygon form the vertices of the new polygon.

Exercise 1.6. *Construct the regular hexagon from the equilateral triangle.*

Exercise 1.7. *Construct the regular octagon from the square.*

The Pythagoreans also knew a construction of the regular pentagon. This construction is more complicated than our previous ones. Figure 1.23 shows a pentagon divided into isosceles triangles with vertex angle of 72°. We show how to construct this angle.

Figure 1.21 illustrates the construction (here, as before, ϕ denotes the golden ratio) and is based on similar triangles. A triangle ABC is constructed with sides of $|AB| = |BC| = \phi$ and $|AC| = 1$. Then $\angle BAC = \angle ACB$. From the defining property of the golden ratio, we have

$$\frac{\phi}{1} = \frac{1}{\phi - 1}.$$

It follows that triangles ABC and ACD are similar (having a common angle and sides on either side of the angle in the same ratio). From this we deduce that $|AD| = 1$, which implies $\angle BAD = \beta$ and $\angle ADC = \alpha$; also $\angle DAC = \beta$ (since $\triangle ABC$ and $\triangle ACD$ have the same angles). Finally, $\alpha = \angle ADC = 2\beta$, as the external angle to $\triangle BAD$. (The implied information is shown in red in figure 1.22.) Thus, from the angle sum in $\triangle ABC$, we have

$$\beta + 2\beta + \alpha = 5\beta = 180°,$$

which yields $\beta = 36°$ and $\alpha = 72°$.

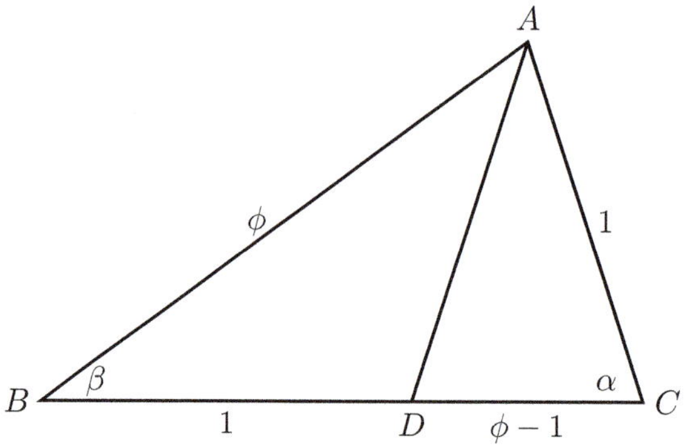

FIGURE 1.21 Construction of a 72° angle

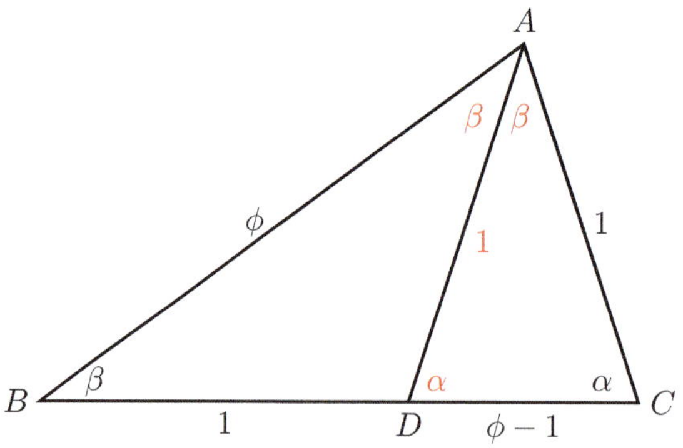

FIGURE 1.22 Additional information

Figure 1.23 depicts the regular pentagon, formed by joining five equally spaced points on the circle, along with the so-called star pentagon or *pentagram*. The Pythagorean school adopted the pentagram as its logo.

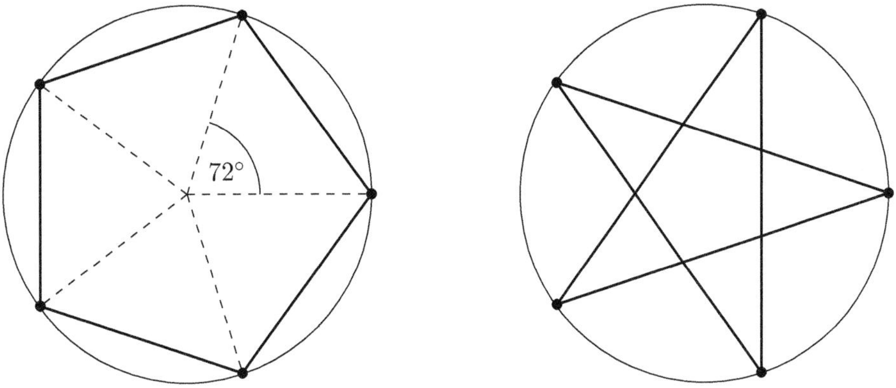

FIGURE 1.23 The regular pentagon and the pentagram

Exercise 1.8. *Consider a pentagram with edges of length one along with three diagonals, as shown in figure 1.24. By considering parallel lines, show the length of line AD is equal to 1.*

Exercise 1.9.* *Continuing from the previous exercise, show using similar triangles that the diagonal length x satisfies $x/1 = 1/(x-1)$.*
(Hint. It might be helpful to look at figure 1.22.)

Exercise 1.10. *Deduce from the previous exercise that the diagonal length of the pentagon is the golden ratio ϕ.*

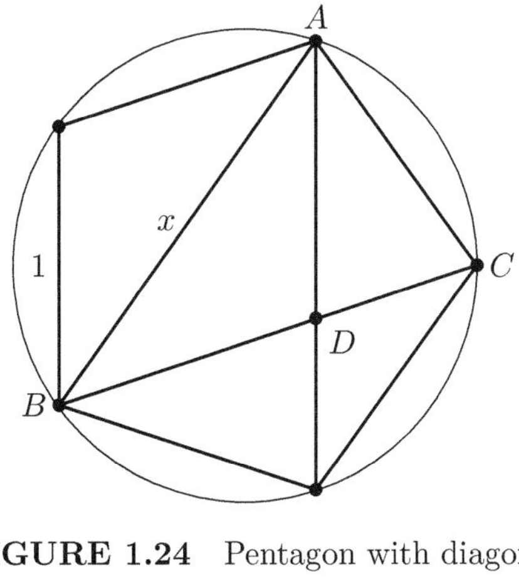

FIGURE 1.24 Pentagon with diagonals

The next regular polygon is the heptagon. The Greeks could not construct this using straightedge and compass. It is not possible, but a proof of this would take several centuries. We will return to this later in the book when we look at some remarkable work by Carl Friedrich Gauss.

1.4 THE WORK OF ARCHIMEDES

FIGURE 1.25 Engraving of Archimedes: Credit: Fjordstone.com

Archimedes was born in the ancient city of Syracuse and lived between c. 287 and c. 212 BCE. He is regarded as the greatest mathematician of antiquity and one of the greatest mathematicians of all time. We don't know much about Archimedes' life except how it ended; legend has it he was killed by a soldier during the Roman invasion of Sicily while drawing circles in the sand. Archimedes had previously designed machines to defend the city, including cranes with huge pincer-like attachments that could (supposedly) pluck ships out of the sea, and massive stone throwers. The invaders must have thought they were under attack from a horde of mechanical demons!

Archimedes was the first to apply mathematical principles to physical phenomena such as the workings of levers and hydrodynamics (behold the famous "Eureka" story). Besides war machines, he invented mechanical devices for peaceful purposes, such as the screw pump and the compound pulley. Despite his innovations in engineering and in applications, Archimedes was most proud of his work in pure mathematics. The great Roman orator and statesman Cicero describes visiting Archimedes' tomb and seeing a sphere surmounted by a cylinder which Archimedes had requested to be placed there to represent his mathematical discoveries.

Trisection of an Angle

Given an arbitrary angle, it is not possible to construct an angle of one-third the size using a straightedge and compass. It is easy to trisect certain specific angles such as a right-angle, but there is no general method that works for all angles. (Pierre Wantzel was the first person to give a proof of this, in 1837.)

For straightedge and compass constructions, we are not allowed to add any marks to the straightedge. It is to be used only for drawing lines between two points and for extending lines. Archimedes showed, however, that if you are allowed to mark your straightedge, trisecting angles becomes possible.

Figure 1.26 shows the construction. The left picture shows the angle α that we want to trisect. We draw it at the center of a circle. We then mark our straightedge with two marks so that the distance between them equal the radius of the circle.

The figure on the right shows the next steps. We extend the straight line containing the diameter of the circle. Then we slide and rotate our straightedge so that it still passes through the point A and so that the distance BC between the circle and the intersection with the extended line equals the radius of the circle. We do this by positioning our straightedge so the one mark is on the circle and the other is on the extended line. This line is shown as ABC in the diagram. The angle this line makes with the extended line, denoted β in the figure is the trisection of α.

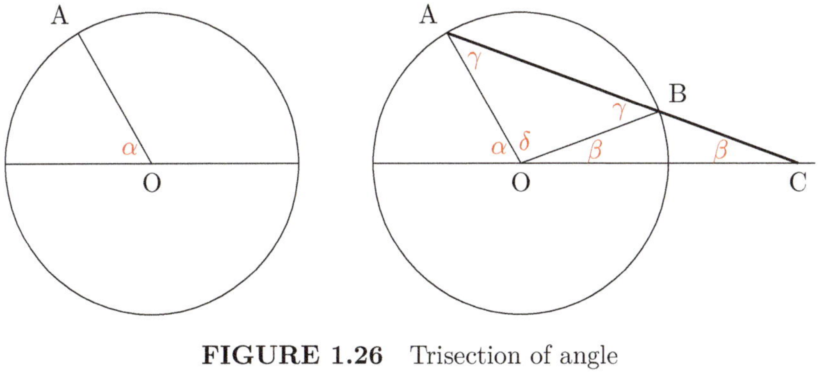

FIGURE 1.26 Trisection of angle

To see why the construction works, observe that the triangles OBC and OAB are both isosceles, so have base angles equal. The base angles of OAB are denoted by γ, but notice that γ is an exterior angle of triangle OBC, so $\gamma = 2\beta$. Since the sum of angles in a triangle equals 180°, we have $\delta = 180-4\beta$. The sum of the angles on the diameter at O sums to 180° since it is a straight line. This gives

$$\alpha + (180 - 4\beta) + \beta = 180,$$

which simplifies to

$$\alpha = 3\beta.$$

Estimation of π

In his treatise *The Measurement of a Circle*, Archimedes gave an estimate for the numerical value of π much more accurate than those of his predecessors. Archimedes' technique was to calculate the areas of a series of regular polygons inscribed within, and circumscribing, the circle. Archimedes used polygons with 6, 12, 24, and 96 sides. The inscribed hexagon (case $n = 6$) is easily constructed. Mark off, starting from any point on the unit circle, chords of radius 1 until the six vertices A, B, C, D, E, and F are obtained, as shown in figure 1.27. These form the vertices of the inscribed hexagon. The vertices G, H, I, J, K, and L of the circumscribed hexagon are found as the intersection of the tangent lines to the circle at A, B, C, D, E, and F. Regular 12, 24-gons, etc. are obtained by successive bisection.

Let p_n and P_n denote the perimeters of the inscribed and circumscribed polygons of n sides. An argument involving the Pythagorean theorem shows that these satisfy the relations

$$P_{2n} = \frac{2p_nP_n}{p_n + P_n}, \quad p_{2n} = \sqrt{p_nP_{2n}}. \tag{1.2}$$

In the case $n = 6$, the figures are regular hexagons, comprised of six equilateral triangles. Clearly, then $p_6 = 6$.

Exercise 1.11. *Show* $P_6 = 4\sqrt{3}$.

Substituting these values into (1.2) results in

$$P_{12} = \frac{48\sqrt{3}}{6 + 4\sqrt{3}}, \quad p_{12} = 12\sqrt{\frac{\sqrt{3}}{3 + 2\sqrt{3}}}.$$

Further substituting in (1.2) gives P_{24} and p_{24}. Combining these values with the then-known estimate

$$\frac{265}{153} < \sqrt{3} < \frac{1351}{780}.$$

Archimedes found the following bounds for π:

$$3\frac{10}{71} < \pi < 3\frac{1}{7}.$$

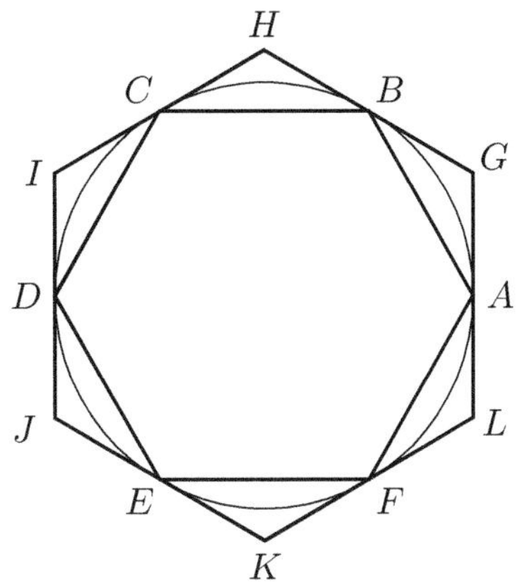

FIGURE 1.27 Inscribed and circumscribed hexagons

Area of the Parabola

The work of Archimedes in this regard is noteworthy because of the use it makes of a powerful method known as *exhaustion*. The method originates with Eudoxus of Cnidos (408–355 BCE, also credited with creating the theory of proportion in Euclid's *Elements*), but came to fruition with the work of Archimedes. Archimedes' argument anticipated the theory of integral calculus.

The underlying idea is to calculate the area of a region (in this case the parabola) by filling, or "exhausting," the region with triangles. Consider the parabolic segment ABC shown in figure 1.28, where C denotes the point on the curve where the tangent line is parallel to the chord AB.[3] Denote the area of $\triangle ABC$ by α. Now inscribe triangles ADC and BCE inside the side segments according to the same recipe. Archimedes proves that each of these two smaller triangles has area $(\frac{1}{8})\alpha$, thus the total area of the triangles is $(\frac{1}{4})\alpha$. Next, construct further triangles with vertices on the parabola and bases on the new chords AD, DC, CE, and EB; these have total area $(\frac{1}{4^2})\alpha$. Continuing this process produces a pattern of triangles whose areas are in geometric progression (each $1/4$ the size of its predecessors). These triangles eventually (after infinitely many steps) exhaust the area of the original parabolic segment.

The combined area of the triangles α_n at the nth stage of the process is

$$\alpha_n = \alpha\Big(\frac{1}{4} + \frac{1}{4^2} + \frac{1}{4^3} + \cdots + \frac{1}{4^n}\Big).$$

[3]Informally, a "tangent line," is a line on a curve that "just touches" the curve at the point of intersection, see figure 1.28 at point C. Tangent lines play an important role in calculus and we define them more precisely in Chapter 5.

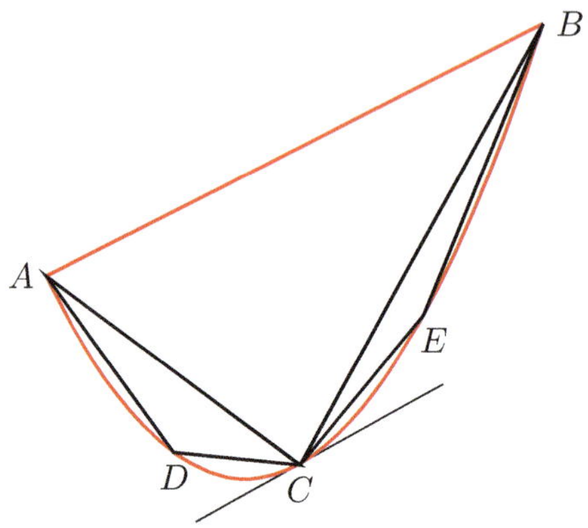

FIGURE 1.28 Area of a parabola

The formula for a geometric sum yields

$$\alpha_n = \frac{4}{3}\alpha\left[1 - \left(\frac{1}{4}\right)^{n+1}\right].$$

Nowadays, in the language of calculus, we would compute the area A of the parabolic segment as the *limit*,

$$\lim_{n\to\infty} \alpha_n = \frac{4}{3}\alpha.$$

But the concept of limit did not exist at that time. Archimedes instead argues by *reductio ad absurdum.* Suppose, on the contrary, that A strictly less than $\frac{4}{3}\alpha$ and write $A = \frac{4}{3}\alpha - \epsilon$, where $\epsilon > 0$. Then we can choose n large enough to make $\alpha_n > A$. By some simple algebra, this is equivalent to choosing n such that

$$\left(\frac{1}{4}\right)^{n+1} < \frac{3}{4}\epsilon.$$

It is clearly possible to do this since ϵ is a fixed positive number and n can be chosen arbitrarily large.[4] This gives a contradiction, since the totality of triangles is contained inside the parabolic segment. The assumption that $A > \frac{4}{3}\alpha$ was shown to lead to a similar contradiction, thereby proving that $A = \frac{4}{3}\alpha$.

In identifying A as $\frac{4}{3}\alpha$ and proving it by this argument, Archimedes came startlingly close to the modern formulation of limit, which did not appear in mathematics until the work of Cauchy in the 1820s. (We use the symbol ϵ in the argument because it is used today in this context.)

[4]This may not have been so clear to the ancients. It relies on the property: given any two positive numbers x and y, there is an integer n such that $nx > y$. This property was introduced in another of Archimedes' works, *On the Sphere and the Cylinder* and is now known as the *Archimedean Axiom*, although these days it is regarded as a proposition rather than an axiom.

Conic Sections and the Archimedean Spiral

Prior to the introduction of analytic geometry in the seventeenth century, the study of curves focused almost exclusively on those arising from conic sections, i.e., formed by the intersection of a cone with a plane. These are the circle, the ellipse, the parabola and the hyperbola, depicted in figure 1.29. The subject of conic sections gained considerable importance in astronomy when it was observed that the paths of heavenly bodies follow these curves, and played a key role in the work of Kepler and Newton.

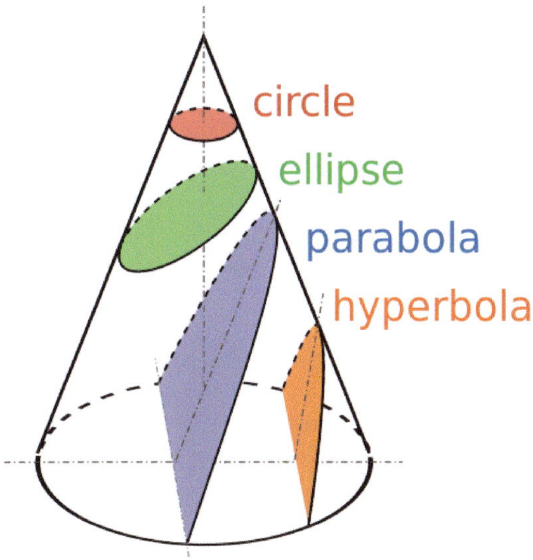

FIGURE 1.29 The conic sections

The most significant advance in the study of conics is due to Apollonius (died 100 BCE), whose eight volumes, aptly titled *Conic Sections*, summarized, and greatly expanded existing work on the subject. In particular, Apollonius proved that a plane cutting a cone would necessarily give rise to a curve of one of the aforementioned types. Apollonius' work was translated into Arabic and had a far-reaching effect on mathematicians of the mid-east.

In this context, we should mention the remarkable discovery by the poet-mathematician Omar Khayyam (1048–1123), of a geometric solution to the cubic equation. Khayyam's solution to a cubic equation, written in the form

$$x^3 + b^2x = b^2c \tag{1.3}$$

is depicted in figure 1.30.

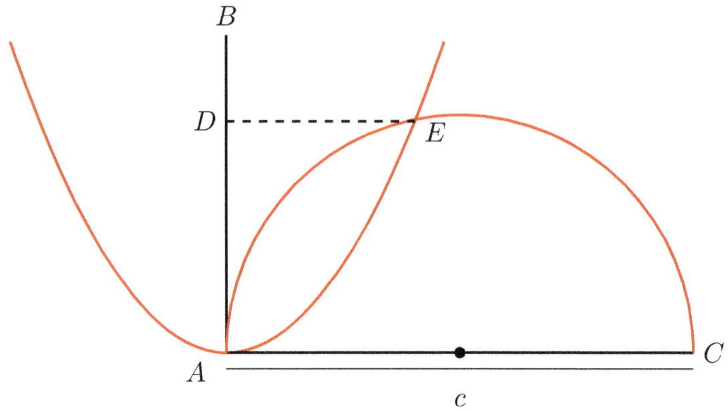

FIGURE 1.30 Omar Khayyam's geometric solution to the cubic equation

Khayyam first constructs the line segment AB, then the parabola with vertex A, axis AB and parameter b.[5] In the language of analytic geometry, which, of course, did not exist at the time, the equation of the parabola would be $by = x^2$, taking point A as the origin. A line segment AC of length c is then constructed perpendicular to AB, and a semicircle is drawn of radius $c/2$ centered at its midpoint. Let E denote the intersection point of the parabola and the semicircle. Then the solution to the cubic equation (1.2) is given by the distance $|DE|$ from the line AC to E.

Exercise 1.12.* *Prove this algebraically.*

Exercise 1.13.* *Prove that equation (1.2) has no more than one (positive) real root.*

In his treatise *On Spirals* Archimedes introduced into mathematics a curve of a new type, now known as the *Archimedean spiral.* The idea is dynamic. In his own words:

If a straight line [half-ray] one extremity of which remains fixed be made to revolve at a uniform rate in the plane until it returns to the position from which it started, and if, at the same time as the straight line is revolving, a point moves at a uniform rate along the straight line, starting from the fixed extremity, the point will describe a spiral in the plane.

The curve is most easily expressed in polar coordinates as $r = a\theta$, where a denotes the rate that the spiral is increasing in radius (see figure 1.31). Among Archimedes' most spectacular achievements is the calculation of the area of the first loop of the spiral ($0 \leq \theta \leq 180°$). Archimedes proved, again using the method of exhaustion, that the area is $\frac{4}{3}\pi^3 a^2$.

[5] A parabola could be constructed in those days as the locus of points equidistant from a fixed point (the *focus*) and a fixed line (the *directrix*).

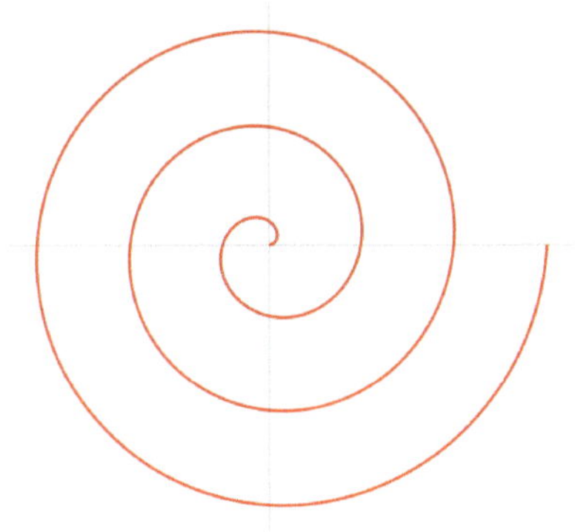

FIGURE 1.31 Archimedean spiral

1.5 PLATONIC SOLIDS

A *Platonic solid* is a convex regular polyhedron in three-dimensional space, where the faces are congruent regular polygons, and where the same number of faces meet at each vertex[6] The most familiar Platonic solid is the cube, or hexahedron. Only four Platonic solids exist besides the cube; they are the *tetrahedron* (with 4 faces), the *octahedron* (8 faces), the *dodecahedron* (12 faces), and the *icosahedron* (20 faces).

The ancient Greeks studied these solids extensively. Some sources credit Pythagoras as their discoverer, although there is some evidence that he may have been unaware of the existence of the octahedron and the icosahedron, and Theaetetus may have discovered them.

The Platonic solids feature frequently in the works of Plato (hence the name), who associated them to the four classical elements, associating *earth* with the cube, *air* with the octahedron, *fire* with the tetrahedron, and *water* with the icosahedron. Perhaps for the sake of inclusivity, Plato remarked that "... God used the dodecahedron for arranging the constellations on the whole heaven." Platonic solids occur in nature, e.g., in the early twentieth century, the biologist Ernst Haeckel discovered a species of radiolaria (a type of protozoa) whose skeletons are shaped like the icosahedron (figure 1.37).

Astronomical references to Platonic solids persisted beyond ancient Greek times. In the sixteenth century, Kepler attempted to relate the positions of the five extraterrestrial planets (the only planets known at that time) to the Platonic solids. In his treatise *Mysterium Cosmographicum*, published in 1596, Kepler constructed a model of the Solar System (figure 1.38) where the five

[6]A *polyhedron* is a solid figure with polygonal faces. *Convex* means that the figure "bulges out" at every point. This is technically described by saying that the straight line joining any two points in the figure is entirely contained within the figure.

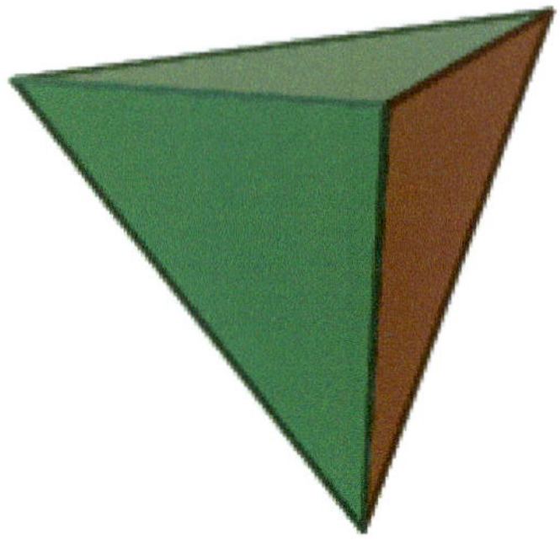

FIGURE 1.32 The tetrahedron

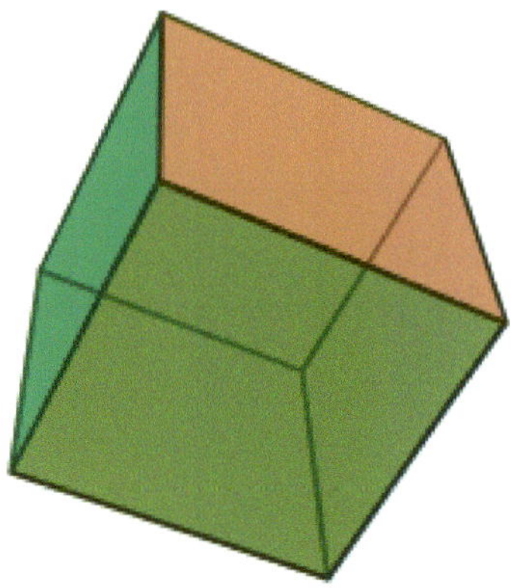

FIGURE 1.33 The hexahedron

Platonic solids were set inside one another and separated by a series of inscribed and circumscribed spheres. Kepler proposed that the distances between the six then-known planets could be understood in terms of the Platonic solids. Kepler eventually abandoned this idea, but out of this work came Kepler's three laws of planetary motion, which greatly influenced Newton's thinking in *Principia Mathematica*. The Platonic solids can be characterized by the pair of integers $\{p, q\}$ giving (respectively) the number of edges of each face, and the number of faces that meet at each vertex. This is known as the Schläfli symbol of the figure. So, for example, the tetrahedron has Schläfli symbol $\{3, 3\}$.

The *Elements* contains a proof that the only existing Platonic solids are the aforementioned ones. The proof, probably originating with Theaetetus, is based on the following observations:

Each vertex V of a Platonic solid must have at least three faces meeting at it. Let m denote the number of faces meeting at V and suppose these faces

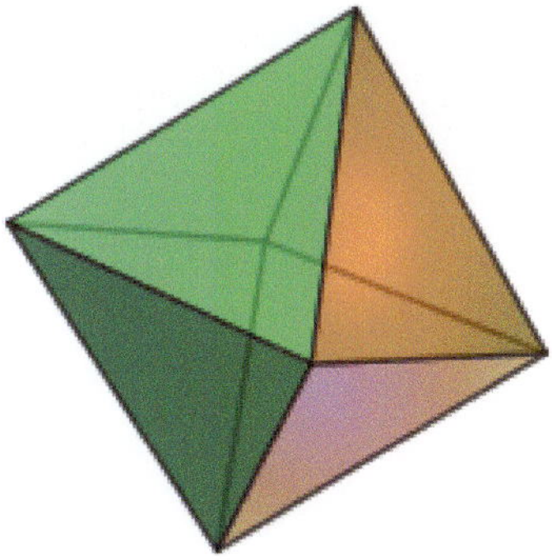

FIGURE 1.34 The Octahedron

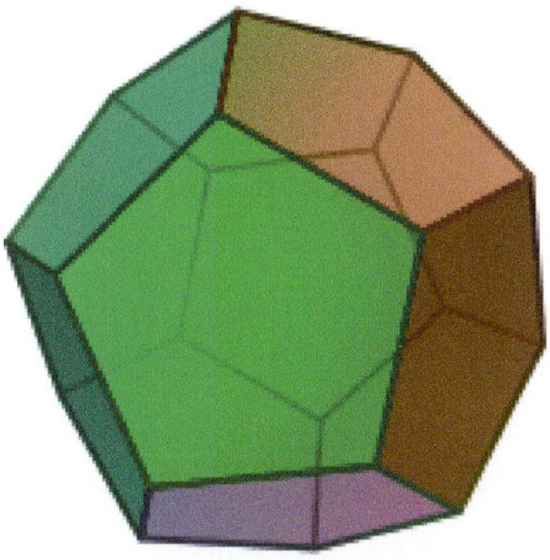

FIGURE 1.35 The Dodecahedron

are regular n-gons with internal angles θ between the sides. It must hold that $m\theta < 360°$. We can see this by unfolding the faces meeting at V of the figure onto the plane containing any one of the faces. The faces then form non-overlapping figures in the plane (see figure 1.39 for an example). This implies the stated inequality.

The following table shows the values of θ corresponding to different values of n:

n	θ
3	60°
4	90°
5	108°
≥ 6	$\geq 120°$

Number of edges versus interior vertex angle

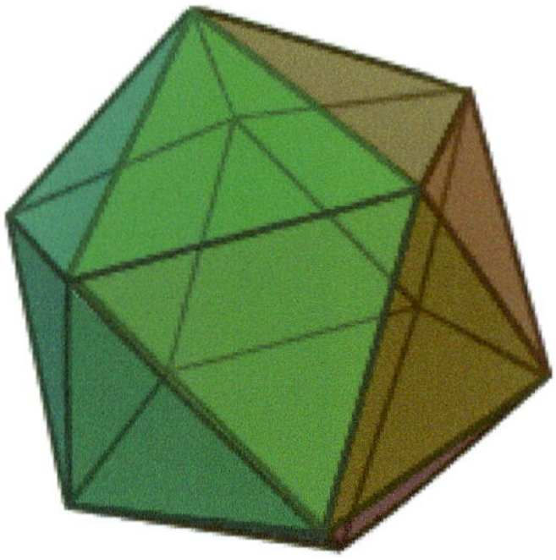

FIGURE 1.36 The icosahedron

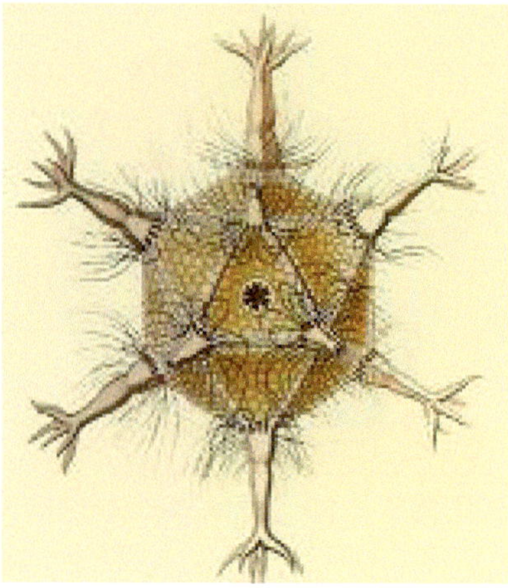

FIGURE 1.37 Circogonia icosahedra, a species of radiolaria

The above considerations strongly limit the feasible range of values of m and n in a Platonic solid. In particular, $n \geq 6$ is not possible, since then the combined angle sum at the vertex would exceed $360°$. In the case $n = 3$, the possible values of m are 3, 4, and 5. These result in tetrahedron, the octahedron, and the icosahedron, respectively. For the case $n = 4$, the only possible choice of m is 3, in which case the figure is a cube. For $n = 5$, again the only possible value for m is 3 and in this case, the figure is the dodecahedron. And these are the only possible Platonic solids!

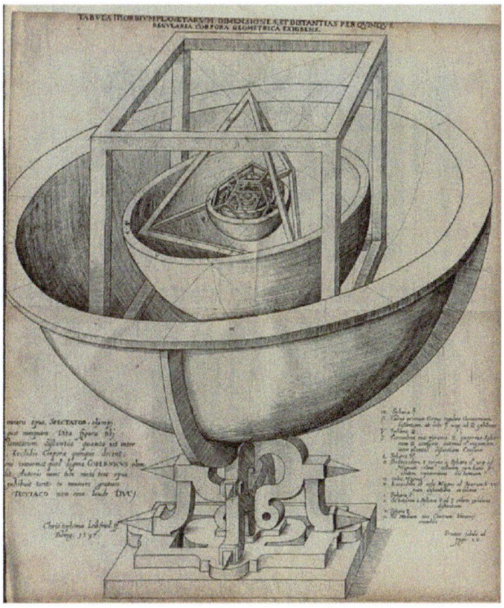

FIGURE 1.38 Kepler's Platonic solid model of the solar system

FIGURE 1.39 Unfolding of the dodecahedron at a vertex

Suggestion for Further Reading

Euclid's Elements. Green Lion Press, 2002.

Bernard Beauzay. *Archimedes' Modern Works.* Société de Calcul Mathématiques SA, 2012.

Sir Thomas L. Heath. *A Manual of Greek Mathematics.* Dover Publications, 2003.

CHAPTER 2

Number Theory

As in the previous chapter, we begin with Euclid. The *Elements* contain, in addition to geometry, some of the fundamental results concerning natural numbers. Here we find proofs that there are an infinite number of prime numbers and that every integer greater than two can be factored into a product of primes.

For the Greeks, numbers were positive integers, but they compared lengths of lines. *Commensurate* lengths correspond to what we now call rational numbers and *incommensurate* lengths to irrational numbers. They proved that the length of a diagonal of the unit square was incommensurate to the length of the side, showing the square root of two is irrational. Their method of comparing lengths of lines is done by an algorithm, and this is where we start the chapter.

We then turn to looking at integer solutions to equations. One important example, is to find Pythagorean triples—the lengths of sides of right triangles. Euclid solves this particular example. Diophantus extended Euclid's work, writing *Arithmetica* consisting of 13 books consisting of methods of finding positive rational solutions to various algebraic problems.

For a more general theory, it helps to work over the integers rather than the natural numbers. We need to extend the definition of *numbers* to include negative numbers and zero. This was done in both China and India. The general solution to finding integer solutions to linear equations comes to us from the work of Brahmagupta.

Our path now takes us from Asia to the Middle East. In ninth century Baghdad, Al-Khwarizmi, much like Euclid, wrote encyclopedic books describing the mathematics known at the time. Some of these books made their way to Europe where they revolutionized mathematics—switching from Roman numerals to the decimal notation makes arithmetic so much simpler!

DOI: 10.1201/9781003470915-2

The study of mathematics in Europe was spurred by translations of Greek and Arabic works. Not all of Diophantus' *Arithmetica* survives, but some does and was translated into Latin. This was the book that inspired Fermat and where he wrote his comments in the margins. We look at both his *Little theorem* and *Last theorem.*

We conclude the chapter by looking at irrational numbers. We start with the Pythagorean's and the proof the square root of two is irrational. Then show that the Euclidean algorithm gives a way of describing numbers using continued fractions. Irrational numbers are either *algebraic* or *transcendental.* We finish with a sketch of Liouville's construction of a transcendental number. The first time a transcendental number was shown to exist.[1]

2.1 THE EUCLIDEAN ALGORITHM

In the *Elements*, Euclid gives a compendium of elementary mathematics known at the time. It is divided into 13 books. The first books are devoted to geometry, but Books VII, VIII, and IX concern numbers. To the Greeks *numbers* meant positive integers, what we call *natural numbers.*

In Book VII, Euclid gives a procedure for finding the greatest common divisor of two natural numbers. This algorithm predates Euclid, but because of his clear exposition and the influence of the *Elements* on succeeding generations of mathematicians, we now call it the *Euclidean Algorithm.*

This algorithm is efficient, and still widely used.

Euclidean Algorithm

Given natural numbers a and b, we let $max(a,b)$ denote the maximum of the two numbers and $min(a,b)$ the minimum.

1. Calculate $max(a,b) - min(a,b)$.
2. If $max(a,b) - min(a,b) = 0$, halt the process—the greatest common divisor is $a = b$.
3. If $max(a,b) - min(a,b) \neq 0$, go to the first step using the two numbers: $min(a,b)$ and $max(a,b) - min(a,b)$.

[1] A few years later Cantor would show that practically all real numbers are transcendental. We will see this in the last chapter of the book.

Example 2.1. *We use the algorithm to find the greatest common divisor of* 114 *and* 42.

$$\begin{aligned}(114, 42) &= \\ (72, 42) &= (114 - 42, 42) \\ (42, 30) &= (42, 72 - 42) \\ (30, 12) &= (30, 42 - 30) \\ (18, 12) &= (30 - 12, 12) \\ (12, 6) &= (12, 18 - 12) \\ (6, 6) &= (12 - 6, 6) \\ &= 6\end{aligned}$$

Why Does the Algorithm Work?

If $max(a, b) - min(a, b) \neq 0$, then $max(a, b) - min(a, b)$ is a natural number that is strictly less than $max(a, b)$. This tells us that the larger of the two numbers decreases at each step of the iteration. There are only a finite number of natural numbers less than $max(a, b)$, so the algorithm must halt after a finite number of iterations.

If d is a divisor of both a and b, then it must also be a divisor of $max(a, b) - min(a, b)$. Conversely, any divisor of both $min(a, b)$ and $max(a, b) - min(a, b)$ must also be a divisor of both a and b. This tells us the greatest common divisor (gcd) of a and b is also the gcd of $min(a, b)$ and $max(a, b) - min(a, b)$. At each stage of the iteration we have two natural numbers, the gcd of each of these pairs are equal. The algorithm halts when we have two equal positive integers. The gcd of two equal positive integers is just the repeated number.

Euclidean Algorithm Using Division

The algorithm as we have presented it uses repeated subtraction. We can speed it up by using division. We use the fact that if m and n are natural numbers, then the remainder r after dividing m by n satisfies $0 \leq r < n$.

Theorem 2.1. *Euclidean Algorithm using division*

1. *Calculate* $max(a, b) - min(a, b)$.

2. *If* $max(a, b) - min(a, b) = 0$, *halt the process—the greatest common divisor is* $a = b$.

3. *If* $max(a, b) - min(a, b) \neq 0$, *calculate the remainder* r *after dividing* $max(a, b)$ *by* $max(a, b) - min(a, b)$.

4. *If* $r = 0$, *halt the process—the greatest common divisor is* $max(a, b) - min(a, b)$.

5. *If* $r \neq 0$, *go to step 1 using the numbers* $max(a, b) - min(a, b)$ *and* r.

Example 2.2. *We repeat our calculation of the gcd of 114 and 42 using division instead of repeated subtraction.*

$$
\begin{array}{ll}
(114, 42) & 114 = 2 \times 42 + 30. \\
(42, 30) & 42 = 30 + 12. \\
(30, 12) & 30 = 2 \times 12 + 6. \\
(12, 6) & 12 = 2 \times 6 + 0.
\end{array}
$$

The Euclidean Algorithm and Number Theory

Euclid's lemma says that if a prime p divides a product ab, then it must divide either a or b. This is an important result, and we will give a proof. Euclid's proof is rather complicated, so we will take a detour and give a simple, more modern proof. Our proof hinges on a result called *Bézout's lemma.* Étienne Bézout was an eighteenth century French mathematician. He proved his lemma holds for polynomials, not just for integers. The result for integers follows easily from the Euclidean algorithm.

Theorem 2.2. *Let* a *and* b *be two natural numbers, then there are two integers* m *and* n *(not necessarily positive) such that*

$$ma + nb = gcd(a, b).$$

Proof. This follows from the Euclidean Algorithm. At each stage, the natural numbers can always be expressed as integer linear combinations of a and b.

□

Example 2.3. *We return to the calculation in example 2.2 but this time keeping track of the linear combinations.*

$$
\begin{aligned}
(114, 42) &= (a, b). \\
(42, 30) &= (b, a - 2b). \\
(30, 12) &= (a - 2b, b - (a - 2b)) = (a - 2b, 3b - a). \\
(12, 6) &= (3b - a, (a - 2b) - 2(3b - a)) = (3b - a, 3a - 8b).
\end{aligned}
$$

The last line tells us that $6 = 3 \times 114 - 8 \times 42$.

Exercise 2.1.*

Let a and b be two natural numbers. Let

$$S = \{ax + by \mid x, y \textit{ are integers}\}.$$

Prove that the smallest positive integer in S is $gcd(a, b)$.

Two integers are called *coprime* if they don't have any primes in common. Equivalently, two integers are *coprime* if their greatest common divisor is 1.

Exercise 2.2. *Suppose that a and b are natural numbers with $gcd(a, b) = d$. Use the previous exercise to show that a/d and b/d are two coprime integers.*

Exercise 2.3. *If a and b are nonzero integers, show $gcd(a, b) = gcd(|a|, |b|)$.*

Using the previous exercise, Theorem 2.2 can be generalized slightly to give the following theorem.

Theorem 2.3. *(Bézout's lemma) Let a and b be two nonzero integers, then there exist two integers m and n such that*

$$ma + nb = gcd(a, b).$$

A *prime number* is an integer greater than 1 whose only divisors are 1 and itself. We can use Bézout's lemma to prove various facts about prime numbers. One that we will use often is Euclid's lemma.

Theorem 2.4. *(Euclid's lemma)*

If p is a prime and it divides the product ab, where a and b are natural numbers, then p divides a or b.

Proof. Suppose that p divides ab but does not divide a. Since the only divisors of p are 1 and p, the gcd of p and a must be 1. Bézout's lemma (theorem 2.2) tells us there must be integers m and n such that

$$ma + np = 1.$$

Multiplying both sides by b gives:

$$mab + npb = b.$$

We know p divides ab, and it clearly divides npb. Since it divides both terms on the left side, it must divide the term on the right side b.

□

Exercise 2.4.* *Show that if p is a prime, and it divides the product abc, where a, b and c are natural numbers, then p divides at least one of a, b, or c.*

Exercise 2.5. *Show that if p is a prime and it divides a product of natural numbers, then it must divide at least one term in the product.*

The Fundamental Theorem of Arithmetic

The Fundamental Theorem of Arithmetic says that every natural number greater than one can be written as a product of primes and this product is unique (up to the ordering of the factors). We split the proof of this into two parts. Given a natural number, we first show how to write it as a product of primes, then we show this product is unique.

Before we start our proof, we comment on the word *product*. In ordinary English, to have a product of terms, you have at least two terms. However, in mathematics, it simplifies things if we extend the definition and allow a product to have just one term.

If we look at the first few natural numbers greater than 1, we have: 2, 3, $4 = 2 \times 2$, 5 and $6 = 2 \times 3$. The prime numbers are considered as products of one term—namely themselves. The uniqueness up to ordering means writing $6 = 2 \times 3$ and $6 = 3 \times 2$ does not give two distinct factorizations—it's the same factorization written with different orderings.

Theorem 2.5. *(Existence of prime factorization)*

We can write very natural number greater than 1 as a product of primes.

Proof. We give a proof by contradiction (*reductio ad absurdum*). We assume that there are natural numbers that cannot be written as a product of primes and deduce a contradiction. Since we cannot have contradictory statements, our initial hypothesis must be false.

If there are natural numbers that don't have prime factorizations, there must be a smallest one. We will call it n. Since n doesn't have a prime factorization, it cannot be a prime. If n is not a prime, we can write $n = ab$, where both a and b are natural numbers smaller than n. Since n is the *smallest* number without a prime factorization, both a and b must have prime factorizations. But this gives us a prime factorization for n—the prime factorization for a concatenated with the prime factorization of b. So n both has and does not have a prime factorization. This is a contradiction telling us that n cannot exist.

We have shown that natural numbers greater than 1 without prime factorizations don't exist or, equivalently, every natural number greater than 1 has a prime factorization.

□

Theorem 2.6. *Uniqueness. Every natural number greater than 1 can be written as a product of primes. This product is unique, up to the order in which the primes are written.*

Proof. Again, we give a proof by contradiction. If there are natural numbers with two (or more) prime factorizations, there must be a smallest one n. The number n has two distinct prime factorizations, which we can write as

$$n = p_1 p_2 \cdots p_k = q_1 q_2 \cdots q_l.$$

Now, the prime p_1 divides the product on the left side, and so must divide the product on the right side. By the result from exercise 2.5, we know that p_1 must divide one of the qs on the right side. This means we can cancel p_1 from both sides, but this gives a number smaller than n with two different prime factorizations. This contradicts the fact that n is the *smallest* natural number with two distinct prime factorizations. Consequently, n cannot exist. All natural numbers greater than 1 have a unique prime factorization.

□

We have split the proof of the Fundamental Theorem of Arithmetic into two parts. The proof of existence appears in Book VII of the *Elements*. Proposition 32 in Book VII says:

> Any number either is prime or is measured by some prime number.

Uniqueness is given by Proposition 14 in Book IX:

> If a number be the least that is measured by prime numbers, it will not be measured by any other prime number except those originally measuring it.

The modern statement and proof of the theorem come from Gauss. He extended the idea of integers to the complex numbers; looking at numbers of the form $a + bi$, where a and b are integers and i is the square root of -1. He showed that, what are now called the *Gaussian integers*, have the unique factorization property. This is not easy. Questions involving factoring become more complicated. For example, 5 is equal to $(1 + 2i)(1 - 2i)$, so it can be factored and is not a Gaussian prime. Gauss' textbook on number theory *Disquisitiones Arithmeticae* (published in 1801) emphasized the importance of the Fundamental Theorem.

Exercise 2.6.* *Later we will need the facts that if m and n are coprime, then mn is coprime to both $m^2 - n^2$ and $m^2 + n^2$. Prove this.*

As an application of the Fundamental Theorem of Arithmetic, we give a proof of the following classical result (another proof will be given in Section 2.12).

Theorem 2.7. *The number $\sqrt{2}$ is irrational, i.e., cannot be expressed as a fraction a/b, where a and b are integers.*

Proof. The proof is by contradiction. Suppose that

$$\sqrt{2} = \frac{a}{b}.$$

Squaring the equation gives

$$a^2 = 2b^2. \tag{2.1}$$

Write $a = 2^m p_1^{r_1} p_2^{r_2} \dots p_k^{r_k}$ and $b = 2^n q_1^{s_1} q_2^{s_2} \dots q_l^{s_l}$ in terms of their prime factorizations, where it assumed that none of the primes $p_1, p_2, \dots, p_k$ nor $q_1, q_2, \dots, q_l$ are 2 (note, it is possible that m or n are 0). Substituting into (2.1), we have

$$2^{2m} p_1^{2r_1} p_2^{2r_2} \dots p_k^{2r_k} = 2^{2n+1} q_1^{2s_1} q_2^{s_2} \dots q_l^{2s_l}.$$

This contradicts the uniqueness part of the Fundamental Theorem of Arithmetic, since the left-hand side has an even power of 2, and the left-hand side an odd power. □

In a similar fashion, it can be proved that whenever $\sqrt{a}$ is not an integer, it is irrational.

Exercise 2.7. *Use this argument to prove that $\sqrt{3}$ is irrational.*

2.2 INFINITUDE OF PRIMES

In Book IX of the *Elements* Euclid gives an elegant proof of a classic theorem of number theory: There are an infinite number of primes. Euclid states this in Proposition 20:

> Prime numbers are more than any assigned multitude of prime numbers.

Our proof of the theorem follows Euclid's argument.

Theorem 2.8. *(Infinitude of primes) There are an infinite number of primes.*

Proof. We show that if we are given any finite collection of primes, we can always find another prime that does not belong to the collection. Suppose we are given primes $p_1, p_2, \dots, p_n$. Define the number n by

$$n = p_1 p_2 \cdots p_n + 1.$$

Dividing n by any of the primes in our collection always gives a remainder of 1, so n is not divisible by any of these primes.

If n is prime, we have found a prime that is not in our original collection. If n is not prime, then it must have a prime factorization. Since all the primes in the factorization of n divide n, none of them can belong to the original collection.

□

2.3 PERFECT NUMBERS

A *perfect number* is one which is equal to the sum of its proper divisors (i.e., all the divisors except the number itself). The definition appeared in the *Elements* (V.22) The first two perfect numbers are

$$6 = 1 + 2 + 3,$$
$$28 = 1 + 2 + 4 + 7 + 14.$$

The early Greek mathematicians knew of two more perfect numbers, 496 and 8128, the last of which was noted by Nicomachus around 100 AD.

We define $\sigma(x)$ to be the sum of all the divisors of x (including x itself), so x is perfect if $\sigma(x) = 2x$.

Exercise 2.8.* *If p and q are coprime, prove*

$$\sigma(pq) = \sigma(p)\sigma(q).$$

Exercise 2.9. *Show $\sigma(2^k) = 2^{k+1} - 1$ for any positive integer k.*

Euclid proved that if $N = 2^p - 1$ is prime, then the number $2^{p-1}N$ is perfect. Prime numbers of this form are named *Mersenne primes*, after the seventeenth century monk Marin Mersenne, who studied them in relation to perfect numbers.

Exercise 2.10.* *Prove that if the number $2^p - 1$ is prime, then p is itself prime.*

In the eighteenth century, Euler proved that this formula yields all even perfect numbers. This result, which we now prove, is known as the *Euclid–Euler theorem*.

Theorem 2.9. *If $N = 2^p - 1$ is prime, then the number $P = 2^{p-1}N$ is perfect. Furthermore, all even perfect numbers have this form.*

Proof. First, assume that $N = 2^p - 1$ is prime. Then the proper divisors of P are

$$1, 2, 2^2, \ldots, 2^{p-1}, N, 2N, 2^2N, \ldots, 2^{p-2}N.$$

Using the formula for the sum of the geometric series, we see that the sum of the proper divisors of P is

$$\begin{aligned}
&\sum_{j=1}^{p-1} 2^j + N \sum_{j=1}^{p-2} 2^j \\
=&(2^p - 1) + N(2^{p-1} - 1) \\
=&N + N(2^{p-1} - 1) \\
=&2^{p-1}N = P,
\end{aligned}$$

showing P is perfect.

Now, suppose that x is an even perfect number. Write x in the form

$$x = 2^k y,$$

where $k \geq 1$ and y is odd.

If x is perfect, then $\sigma(x) = 2x$ and using and exercises 2.8 and 2.9, we have

$$2^{k+1} y = \sigma(2x) = \sigma(2^k)\sigma(y) = (2^{k+1} - 1)\sigma(y). \tag{2.2}$$

This implies that the odd number $2^{k+1} - 1$ divides y, i.e.,

$$\frac{y}{2^{k+1} - 1}$$

is a proper factor of y.

Let Q denote the sum of any other proper factors of y. Then

$$\begin{aligned}\sigma(y) &= y + \frac{y}{2^{k+1} - 1} + Q \\ &= \frac{2^{k+1} y}{2^{k+1} - 1} + Q.\end{aligned}$$

Together with 2.2, this implies $Q = 0$. Thus y is prime and

$$\frac{y}{2^{k+1} - 1} = 1.$$

This proves that x has the claimed form. □

The Euler–Euclid theorem thus characterizes all even perfect numbers. They are in one-to-one correspondence with the Mersenne primes. The first four primes, $p = 2, 3, 5$, and 7 yield $N = 2^p - 1 = 3, 7, 31$, and 127. These are all primes and give rise, by the Euler–Euclid recipe, to the perfect numbers 6, 28, 496, and 8128. The next possible candidate for a Mersenne prime $2^{11} - 1 = 2047$ is not actually prime, having the factorization 23×89.

The Mersenne numbers $2^p - 1$ turn out to be prime for only 43 of the first two million primes p. It is not known if infinitely many Mersenne primes, and hence infinitely many even perfect numbers, exist.

It is likewise unknown to this day whether or not any odd perfect numbers exist.

2.4 DIOPHANTINE EQUATIONS

A Diophantine equation is an equation for which we are only interested in integer solutions. An important early example involves Pythagorean triples: What are the natural number solutions to $x^2 + y^2 = z^2$?

These equations are named after Diophantus, a Greek mathematician who lived after Euclid. Diophantus wrote *Arithmetica*, consisting of 13 books on methods for finding integer and rational solutions to equations. Not all of the 13 books survived, but the portion that did was translated from the Greek into Latin in 1570. Pierre de Fermat, who we will meet later owned and studied from a copy. This is the book in which he wrote that he had a solution to his "Last Theorem" but the margin was too small to contain it.

Diophantus lived in Alexandria, but not much else is known about him. His name appears in a problem written by a much later Greek author, Metrodorus.

> "Here lies Diophantus," the wonder behold.
> Through art algebraic, the stone tells how old:
> "God gave him his boyhood one sixth of his life,
> One twelfth more as youth while whiskers grew rife;
> And yet one-seventh ere marriage begun;
> In five years there came a bouncing new son.
> Alas. the dear child of master and sage
> Attained only half of his father's full age.
> When chill fate took him — an event full of tears —
> Heartbroken, his father lived just four more years."

If we let x denote the length of Diophantus' life, we obtain the equation:

$$x = x/6 + x/12 + x/7 + 5 + x/2 + 4.$$

Solving gives $x = 84$.

2.5 PYTHAGOREAN TRIPLES

A *Pythagorean triple* consists of three natural numbers: a, b, and c such that

$$a^2 + b^2 = c^2.$$

These triples correspond to right triangles whose sides have integer lengths. As we commented in the previous chapter, the Pythagorean theorem predates Pythagoras. A clay tablet (Plimpton 322) dating back to 1800 BC contains a list of Pythagorean triples.

Two commonly known triples are $(3, 4, 5)$ and $(5, 12, 13)$. Once you have a triple, you can form another triple by multiplying a, b, and c by a common factor. For example, starting with the triple $(3, 4, 5)$, we know $(6, 8, 10)$, $(9, 12, 15)$ and so on, will also be Pythagorean triples. It is easily checked that if two of a, b or c have a common factor, then the third number must also have this factor.

Primitive Pythagorean triples are Pythagorean triples where $gcd(a, b) = gcd(a, c) = gcd(b, c) = 1$. Every other Pythagorean triple can be obtained by multiplying a primitive triple by a common factor. We would like a way to find the primitive triples.

Both Euclid and Diophantus had methods for generating primitive Pythagorean triples. We look at both.

Euclid's formula

Algebra using symbols for variables wasn't invented until long after Euclid's life, so he didn't express the formula as we would nowadays. He gave a construction using lengths of lines. If we assign letters to the lengths, his construction gives what we now call Euclid's formula.

Theorem 2.10. *(Euclid's formula) Given any primitive Pythagorean triple* (a, b, c)*, where* a *is odd, we can find coprime natural numbers* m *and* n*, one of which is even, such that* $a = m^2 - n^2$*,* $b = 2mn$*, and* $c = m^2 + n^2$*.*

Conversely, given two coprime natural numbers, m *and* n*, one of which is even, if we define* a*,* b*, and* c *by* $a = m^2 - n^2$*,* $b = 2mn$*, and* $c = m^2 + n^2$*, then* (a, b, c) *is a primitive Pythagorean triple.*

Notice that if we have a Pythagorean triple, $a^2 + b^2 = c^2$, with both a and b even, then c must be even. So, if we have a primitive Pythagorean triple, at least one of a and b must be odd. We can always relabel the variables to make a odd.

Before we start the proof, we need one fact that we leave as an exercise.

Exercise 2.11. *Show that if* m *and* n *are two odd integers, then* $m^2 - n^2$ *is divisible by 4.*

Proof. Let (a, b, c) be a primitive Pythagorean triple with a odd. We have

$$a^2 + b^2 = c^2.$$

Rearranging gives:

$$c^2 - a^2 = b^2.$$

We can factor the left side of the equation to obtain:

$$(c - a)(c + a) = b^2.$$

More rearranging gives:

$$\frac{c + a}{b} = \frac{b}{c - a}. \tag{2.3}$$

Reduce the fraction on the left side to lowest terms. Let m be the numerator and n the denominator of this reduced fraction. We have

$$\frac{c + a}{b} = \frac{m}{n}, \tag{2.4}$$

with m and n coprime. From equation 2.3, we obtain

$$\frac{b}{c-a} = \frac{m}{n}, \tag{2.5}$$

which gives

$$\frac{c-a}{b} = \frac{n}{m}. \tag{2.6}$$

Rewriting equations 2.4 and 2.6, we obtain the system of two equations:

$$\frac{c}{b} + \frac{a}{b} = \frac{m}{n} \tag{2.7}$$
$$\frac{c}{b} - \frac{a}{b} = \frac{n}{m}.$$

Adding the two equations and dividing by 2 gives:

$$\frac{c}{b} = \frac{m^2+n^2}{2mn}. \tag{2.8}$$

Subtracting the second equation from the first and dividing by 2 gives:

$$\frac{a}{b} = \frac{m^2-n^2}{2mn}. \tag{2.9}$$

Since (a, b, c) is a primitive Pythagorean triple, a and b are coprime. Consequently, a/b is a reduced fraction.

We will now show $(m^2 - n^2)/2mn$ is also a reduced fraction. Since m and n are coprime we know by exercise 2.6 that $m^2 - n^2$ and mn are coprime. The only possibility for $(m^2 - n^2)/2mn$ to not be reduced is if we can cancel 2 from both the numerator and denominator.

However, since m and n are coprime, the only way for the numerator to be even is if both m and n are odd. But if they are both odd, exercise 2.11 tells us the numerator will still be even after dividing by 2. This says that in reduced form the numerator is even, but we know that in reduced form it equals a/b and a is odd, so this cannot happen. One of m and n must be even and the fraction $(m^2 - n^2)/2mn$ is in reduced form.

Since

$$\frac{a}{b} = \frac{m^2-n^2}{2mn} \tag{2.10}$$

and both fractions are in reduced form, we have $a = m^2 - n^2$ and $b = 2mn$. Equation 2.8 shows that $c = m^2 + n^2$. This completes the proof of the first part, we must now prove the converse.

Suppose m and n are two coprime natural numbers, one of which is even. We define a, b, and c by $a = m^2 - n^2$, $b = 2mn$, and $c = m^2 + n^2$. We obtain:

$$a^2 + b^2 = (m^2 - n^2)^2 + 4m^2n^2 = m^4 + 2m^2n^2 + n^4 = (m^2 + n^2)^2 = c^2.$$

So (a, b, c) is a Pythagorean triple. Exercise 2.6 shows it is primitive. □

Diophantus' Method

In the *Arithmetica*, Diophantus asks how the square of a rational number can be expressed as the sum of two other rational numbers. In other words, he is looking for rational solutions for $a^2 + b^2 = c^2$. He shows how to solve this in one case. In Fermat's copy of the book, it is the margin by this problem where he wrote:

> It is impossible to separate a cube into two cubes, or a fourth power into two fourth powers, or in general, any power higher than the second, into two like powers. I have discovered a truly marvelous proof of this, which this margin is too narrow to contain.

Diophantus' solution gives us a way of constructing Pythagorean triples. This method is sometimes called the "Diophantus chord" method.

Given a Pythagorean triple (a, b, c), we know that $a^2 + b^2 = c^2$. Dividing by c^2 gives:

$$\left(\frac{a}{c}\right)^2 + \left(\frac{b}{c}\right)^2 = 1.$$

If we consider the circle with equation $x^2 + y^2 = 1$, the Pythagorean triple gives us a point on this circle with rational coordinates. Conversely, given two rational numbers m/n and r/n that satisfy

$$\left(\frac{m}{n}\right)^2 + \left(\frac{r}{s}\right)^2 = 1,$$

we can multiply both sides by the denominators to obtain $(ms)^2 + (rn)^2 = (ns)^2$, telling us (ms, rn, ns) is a Pythagorean triple. The question of finding Pythagorean triples is equivalent to finding all points on the circle $x^2+y^2 = 1$ with rational coefficients.

Consider the unit circle with a ray drawn from the point with coordinates $(-1, 0)$ to some other point P on the circle. This is shown in figure 2.1.

Let μ denote the slope of the secant line. The line has equation $y = \mu(x+1)$. (We usually use m for slopes of lines, but in this chapter m is usually used for an integer. We use the Greek letter μ to indicate that the slope need not be rational.) The intersection of the line and the circle can be found by plugging $y = \mu(x + 1)$ into the equation $x^2 + y^2 = 1$. this gives:

$$x^2 + [\mu(x + 1)]^2 = 1.$$

Expanding and rearranging gives the following quadratic equation in x:

$$(\mu^2 + 1)x^2 + 2\mu^2 x + \mu^2 - 1 = 0. \tag{2.11}$$

We know that $(x + 1)$ is a factor. Equation 2.11 factors:

$$(x + 1)\left((\mu^2 + 1)x + (\mu^2 - 1)\right) = 0,$$

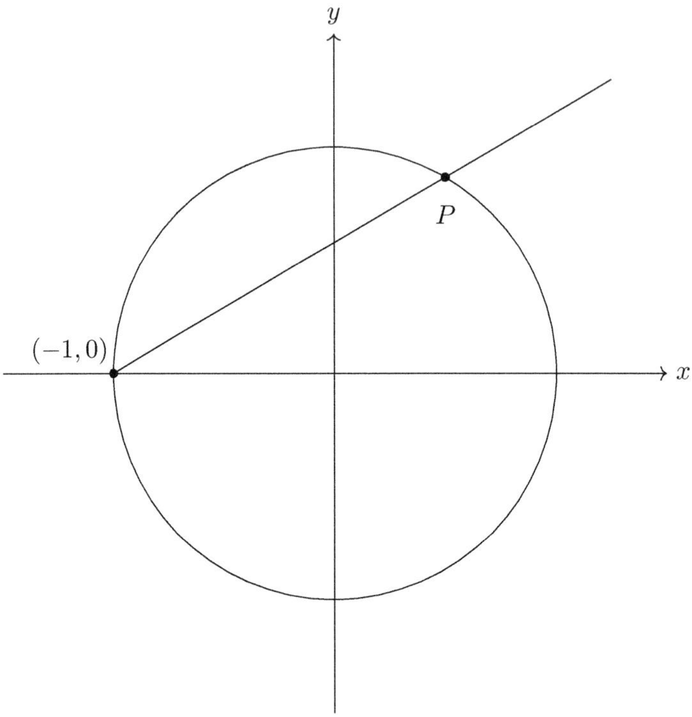

FIGURE 2.1 Unit circle with secant line

which gives

$$x = \frac{1 - \mu^2}{\mu^2 + 1}$$

as the x coordinate of P. Plugging this value back into the equation of the line gives the y coordinate

$$y = \frac{2\mu}{\mu^2 + 1}.$$

It is straightforward to check that if μ is rational, then both the x and y coordinates must be rational. It is also clear that if x and y are rational, then the slope of the line connecting the point $(-1, 0)$ to (x, y) must be rational. This tells us the points on the unit circle with rational coefficients, except for $(-1, 0)$, have the form

$$\left(\frac{1 - \mu^2}{\mu^2 + 1}, \frac{2\mu}{\mu^2 + 1} \right),$$

where μ runs through the rational numbers.

We are really interested in Pythagorean triples consisting of natural numbers. These correspond to choosing the point P, shown in figure 2.1, in the first quadrant. This means we want μ to satisfy $0 < \mu < 1$. Since μ is rational, we can write it as n/m where both n and m are natural numbers.

Putting $\mu = n/m$ in the formula for the x coordinate gives:

$$x = \frac{1 - (n/m)^2}{(n/m)^2 + 1}.$$

Multiplying both the numerator and denominator by n^2 yields

$$x = \frac{m^2 - n^2}{m^2 + n^2}.$$

A similar calculation gives

$$y = \frac{2mn}{m^2 + n^2}.$$

These two expressions give us Euclid's formula.

Diophantus' chord method of finding points on the circle with rational coefficients can be extended to rational solutions to other certain other quadratic equations, for example, functions of the form $x^2 - ny^2 = 1$, where n is a non-square integer.

Though finding rational solutions to certain quadratic equations can be easy, finding integer solutions is much harder. Instead of quadratics, we start by considering the simpler case of linear functions.

2.6 LINEAR DIOPHANTINE EQUATIONS

The general solution for a linear Diophantine equation in two variables was given by Brahmagupta in his *Brāhmasphuṭasiddhānta* (*Correctly established doctrine of Brahma*), written in 628 CE. initially, it might seem surprising that Greek mathematicians didn't prove this centuries before Brahmagupta, but to give a complete solution you need to consider negative numbers and zero as legitimate numbers. For the Greek mathematicians, numbers were positive. Extending the natural numbers to include negative numbers was first done in China and then in India. After this, Indian mathematicians included zero as a number. Brahmagupta's *Brāhmasphuṭasiddhānta* contains a number of results concerning integers.

Theorem 2.11. *Let a and b be nonzero integers with $d = gcd(a, b)$. For any integer c, the equation $ax + by = c$ has a solution if and only if c is divisible by d.*

Proof. We know d divides both a and b, so it must divide $ax + by$ for any integers x and y. So, if $ax + by = c$ has a solution in integers, d must divide c.

Conversely, we must show if c is divisible by d, then there is a solution. Let $c = kd$ for some integer k. We know by Theorem 2.3, there is a solution to $ax + by = d$. This means we can find integers m and n such that $am + bn = d$. Clearly, $a(km) + b(kn) = kd = c$, so $ax + by = c$ has a solution: $x = km$, $y = kn$. □

Given a Diophantine equation $ax + by = c$ with c is divisible by $gcd(a, b)$, we know how to find one solution using the Euclidean algorithm. There are infinitely many other solutions. The following theorem gives all the solutions.

Theorem 2.12. *(Brahmagupta) Let a and b be nonzero integers with $d = gcd(a, b)$. Let $x = m$ and $y = n$ be a solution to the Diophantine equation $ax + by = c$.*

For any integer t, $x = m + (b/d)t$, $y = n - (a/d)t$ is also a solution.

Conversely, every solution to $ax + by = c$ can be written as $x = m + (b/d)t$, $y = n - (a/d)t$ for some integer t.

Proof. First, we show $x = m + (b/d)t$, $y = n - (a/d)t$ is a solution. Plugging these values into $ax + by$ gives:

$$a(m + (b/d)t) + b(n - (a/d)t) = am + bn = c.$$

So, we have a solution.

For the converse, assume $x = r$, $y = s$ is another solution. We have $am + bn = c$ and $ar + bs = c$. Subtracting the second equation form the first gives:

$$a(m - r) + b(n - s) = 0. \tag{2.12}$$

Rearranging gives:

$$a(m - r) = -b(n - s).$$

Dividing by the greatest common divisor:

$$(a/d)(m - r) = -(b/d)(n - s).$$

Now the integer a/d divides the left side of the equation and so must divide the right. We know by exercise 2.2 that $gcd(a/d, b/d) = 1$, so a/d shares no prime factors with b/d, and must divide $n - s$. This means there must be some integer t with

$$n - s = (a/d)t.$$

Rearranging, we obtain

$$s = n - (a/d)t.$$

To obtain the value for r we plug $s = n - (a/d)t$ into equation (2.1) to obtain

$$a(m - r) + b(a/d)t = 0.$$

Dividing by a and solving for r gives:

$$r = m + (a/d)t.$$

□

We give an example of a typical type of problem that can be solved using the previous theorem.

Example 2.4. *Domestic stamps cost 42 cents and international cost \$1.14. I go to the post office and buy several stamps. The total comes to \$13.20. How many of each type did I buy?*

Let x denote the number of international stamps and y the number of domestic stamps. Expressing the prices in cents gives the equation:

$$114x + 42y = 1320.$$

In example 2.2, we used the Euclidean algorithm to show $gcd(114, 42) = 6$. In example 2.3 we used the steps of the calculation to show $114 \times 3 - 42 \times 8 = 6$.

This tells us that $x = 3$, $y = -8$ is a solution to $114x+42y = 6$. Multiplying by 220 tells us $x = 660$, $y = -1760$ is a solution to $114x - 42y = 18$.

Theorem 2.12 then says that every integer solution has the form:

$$x = 660 + 7t, y = -1760 - 19t,$$

where t can be any integer.

We are only interested in non-negative integer solutions, so we must have

$$660 + 7t \geq 0 \quad \textit{and} \quad -1760 - 19t \geq 0.$$

This gives:

$$-660/7 \leq t \leq -1760/19$$

or

$$-94 \leq t \leq -93.$$

The question has two possible answers. When $t = -94$, we obtain $x = 2$ $y = 26$. When $t = -93$, $x = 9$, $y = 7$.

2.7 AL-KHWARIZMI

Muhammad ibn Mūsā al-Khwarizmi was a Persian mathematician and astronomer and the head of the "House of Wisdom" in Baghdad during the early ninth century. "The House of Wisdom," also known as the "Grand Library of Baghdad" housed a collection of work from Greek, Persian and Indian sources. Scholars studied mathematics, science and philosophy there.

FIGURE 2.2 Muhammad ibn Mūsā al-Khwarizmi

Al-Khwarizmi wrote two books that transformed mathematics in Europe. The first, written around 830 CE, was *Al-Kitāb al-Mukhtaṣar fī Ḥisāb al-Jabr wa-Muqābalah* (*The Compendious Book on Calculation by Completion and Balancing*) was a popular book on methods of solving quadratic equations designed for a wide audience. It was translated into Latin by Robert of Chester in 1145 CE. His translation of the title is *Liber Algebrae et Almucabola*. The Latinization of *al-Jabr* became *algebrae* which in turn gives us the word *algebra*.

When we use the word *algebra* today, we usually mean symbolic algebra, where symbols like x and y represent unknown quantities, but this is not the meaning of the word when we describe early mathematics. Diophantus and al-Khwarizmi are often described as inventing algebra because they were giving methods for solving equations. These methods were described using words, not symbols. Indian mathematics was often written in verse. (Symbolic algebra began with François Viète in the sixteenth century.)

Al-Khwarizmi also wrote about the Indian decimal place system for writing numbers. He explained how to do addition and subtraction using this notation. His original writings on the decimal system no longer exist, but there is Latin translation *Algoritmi de numero Indorum* from the twelfth century. Calculations using the decimal place system are much simpler than using Roman numerals and, not surprisingly, Europe adopted the new system. The title of al-Khwarizmi's work clearly acknowledges the Indian origin, but his name became the one associated with it in the west. Algoritmi is the Latinization of al-Khwarizmi, the word *algorist* became associated with people who performed calculations using the decimal system. Over time, *algorist* became *algorithm* and the meaning changed to its current one of being a set of rules to perform a computation.

Although Arab mathematicians[2] had access to Indian mathematics, they did not accept negative numbers and zero as being proper numbers. As with the Greeks, numbers meant positive numbers. Consequently, al-Khwarizmi's work only used positive numbers. In Europe, it would take several centuries before negative numbers were fully accepted.

2.8 NUMBER THEORY IN EUROPE

Pierre de Fermat (1607–1665) studied law and then practiced it in the *Parlement* of Toulouse, where he worked his whole life. In his spare time, he began by looking at Greek mathematical works and from there, he began to generate his own results. His work is known through notes he made and correspondence with other mathematicians. He enjoyed generating new ideas, but was not good at explication, and published very little. He developed an algebraic approach to geometry which probably predates Descartes', but was published after his death. His ideas on tangent lines and finding areas under curves helped in the development of calculus. He also worked in number theory.

Fermat stated many results concerning numbers. Unfortunately, he rarely gave the proofs and number theory was not a focus of mathematicians at the time. After his death, his son Samuel collected his correspondence and published it along with the copy of *Arithmetica* with Fermat's marginal notes. Leonhard Euler (1707–1783) took Fermat's results, provided proofs where he could, and generalized the ideas. Euler's work made number theory an important area of study once more.

We will look at some of Fermat's and Euler's work, but first introduce modular arithmetic. Gauss introduced this to help simplify ideas concerning remainders and provide an elegant way of thinking about number theoretic problems.

[2] *Arab mathematicians* refers to mathematicians who wrote in Arabic. This is similar to *Greek mathematicians* referring to mathematicians who wrote in Greek.

FIGURE 2.3 Pierre de Fermat

2.9 MODULAR ARITHMETIC

Let n be a positive integer. Given integers a and b, we say *a is congruent to b modulo n* if n divides $b - a$. If a is congruent to b modulo n, write

$$a \equiv b \mod (n).$$

This is a succinct notation telling us that the remainder of a divided by n is equal to the remainder of b divided by n.

As an example we will take $n = 5$. When we divide by 5 there are 5 possible outcomes: we get a remainder of 0, 1 , 2, 3, or 4. This divides the integers into five disjoint subsets:

$$\{\dots, -10, -5, 0, 5, 10, \dots\}$$
$$\{\dots, -9, -4, 1, 6, 11, \dots\}$$
$$\{\dots, -8, -3, 2, 7, 12, \dots\}$$
$$\{\dots, -7, -2, 3, 8, 13, \dots\}$$
$$\{\dots, -6, -1, 4, 9, 14, \dots\}$$

These subsets of the integers are called the *equivalence classes modulo* 5.

In general, there are n equivalence classes modulo n. Every integer belongs to one and only one of these classes. We choose one number as a *representative* for its class. Usually, we pick the representatives to be $0, 1, \dots, n - 1$.

Exercise 2.12.* *Prove that if* $a \equiv b \mod (n)$ *and* $c \equiv d \mod (n)$*, then* $a + c \equiv b + d \mod (n)$.

Exercise 2.13. *Prove that if* $a \equiv b \mod (n)$ *and* $c \equiv d \mod (n)$*, then* $a - c \equiv b - d \mod (n)$.

Exercise 2.14. *Prove that if* $a \equiv b \mod (n)$ *and* $c \equiv d \mod (n)$*, then* $ac \equiv bd \mod (n)$.

Exercises 2.12, 2.13, and 2.14 tells us we can add a number to both sides of a modular equation, subtract a number from both sides, multiply both sides by a number, and keep the congruence. Using these operations we can solve some congruences in exactly the same way as we do for regular equations.

Suppose we are given the congruence $x+2 \equiv 5 \mod (7)$ and want to solve for x. We can subtract 2 from both sides to obtain $x \equiv 3 \mod (7)$. However, because we are working over the integers, we cannot, in general, use division. For example, if we have

$$2x \equiv 3 \mod (7), \tag{2.13}$$

we cannot divide by 2 because 3/2 is not an integer. However, the congruence 2.13 does have a solution, namely $x \equiv 5 \mod (7)$. To obtain solutions to $ax \equiv b \mod (n)$ we can restate some of our previous theorems using modular notation.

Solving $ax \equiv b \mod (n)$

To solve $ax \equiv b \mod (n)$ we need to find the values of x that make $ax - b$ divisible by n. This is equivalent to solving

$$ax - b = ny,$$

or equivalently

$$ax - ny = b.$$

We know how to do this. We can restate Theorem 2.11 to give theorem.

Theorem 2.13. *The congruence* $ax \equiv b \mod (n)$ *has a solution if and only if the* $gcd(a, n)$ *divides* b.

We can restate Brahmagupta's theorem (Theorem 2.12) to give all solutions.

Theorem 2.14. *If* $x = m$ *is a solution to* $ax \equiv b \mod (n)$*, then all other solutions are given by* $x = m + (n/d)t$ *where* t *runs through the integers and* d *denotes* $gcd(a, n)$.

We can restate the previous two results using equivalence classes:

Theorem 2.15. *Let $d = gcd(a, n)$. The congruence*

$$ax \equiv b \mod (n)$$

has a solution if and only if d divides b. Suppose d divides b and m is a solution. Then the solutions form d congruence classes, with representatives

$$m, m + \frac{n}{d}, m + \frac{2n}{d}, \ldots, m + \frac{(d-1)n}{d}.$$

As an immediate consequence of the above theorem we have $ax \equiv 1 \mod (n)$ has a solution if and only if $gcd(a, n) = 1$. We say that $a \mod (n)$ has a *multiplicative inverse mod(n)* if we can find b such that $ab \equiv 1 \mod (n)$.

Corollary 2.1. *An integer a has a multiplicative inverse modulo(n) if and only if a and n are coprime. If a has a multiplicative inverse modulo(n), it is unique modulo(n).*

Having a multiplicative inverse can be useful when simplifying expressions. For example, suppose we are given

$$3x \equiv 3 \mod (5).$$

Since 3 and 5 are coprime, we know 3 has a multiplicative inverse. Multiplying both sides by this inverse gives

$$x \equiv 1 \mod (5).$$

If instead, we were given

$$3x \equiv 3 \mod (12),$$

we cannot cancel the 3s, because 3 and 12 are not coprime, and so 3 does not have a multiplicative inverse. To solve, we use Theorem 2.15 to obtain the three solutions:

$$x \equiv 1, x \equiv 5, x \equiv 9 \mod (12).$$

We will use Corollary 2.1 several times. It helps to have it expressed slightly differently.

Corollary 2.2. *If a and n are coprime, and $ax \equiv ay \mod (n)$, then $x \equiv y \mod (n)$.*

Exercise 2.15.* *If a and n be coprime, show $0, a, 2a, \ldots, (n-1)a$ belong to distinct congruence classes modulo n. This shows that multiplying by a gives a permutation of the congruence classes.*

Exercise 2.16. *If a and n be coprime, show $a, 2a, \ldots, (n-1)a$ belong to distinct non-zero congruence classes modulo n. This shows that multiplying by a gives a permutation of the non-zero congruence classes.*

2.10 FERMAT'S LITTLE THEOREM

In one of Fermat's letters, he states that if p is a prime and a is an integer not divisible by p, then $a^{p-1} - 1$ is divisible by p. This result is now known as his *little theorem.* If Fermat had a proof, it hasn't been found. The first published proof was by Euler in 1736.

Theorem 2.16. *(Fermat's little theorem) If p is a prime and $a \not\equiv 0 \mod (p)$, then $a^{p-1} \equiv 1 \mod (p)$.*

Proof. We know a and p are coprime. Exercise 2.16 tells us that the non-zero congruence classes are permuted by multiplying by a. We can write the product of the non-zero congruence classes as either

$$1 \times 2 \times 3 \times \cdots \times (p-1) \mod (p)$$

or

$$a \times 2a \times 3a \times \cdots \times (p-1)a \mod (p).$$

Since these are equal

$$1 \times 2 \times 3 \times \cdots \times (p-1) \equiv a \times 2a \times 3a \times \cdots \times (p-1)a \mod (p).$$

We can simplify:

$$1 \times 2 \times 3 \times \cdots \times (p-1) \equiv 1 \times 2 \times 3 \times \cdots \times (p-1)a^{p-1} \mod (p).$$

Notice that $2, 3, \ldots, p-1$ are all coprime to p. Corollary 2.2 tells us that we can cancel these from both sides, giving

$$1 \equiv a^{p-1} \mod (p).$$

□

Euler wanted to generalize this theorem for cases when n is not a prime. The proof we have just given hinges on the fact that $2, 3, \ldots, p-1$ are all coprime to p. If n is not a prime, then it is not true that $2, 3, \ldots, n-1$ are all coprime to n.

Exercise 2.17.[*] *If $n > 4$ is composite, show $1 \times 2 \times 3 \times \cdots \times (n-1) \equiv 0 \mod (n)$.*

Instead of looking at all the congruence classes we can look at the congruence classes containing integers that are coprime to n. For example, if $n = 12$ representatives of these classes are 1, 5, 7, and 11. Euler denoted the number of classes by $\phi(n)$—this function is now called the *Euler phi function*—so $\phi(12) = 4$.

Exercise 2.18. *If p is a prime, show $\phi(p) = p - 1$.*

Exercise 2.19. *Show that if a and b are both coprime to n, then ab is coprime to n.*

Exercise 2.20. *Let $\{a_1, a_2, \ldots, a_{\phi(n)}\}$ be a set of representatives of the congruence classes coprime to n. Let a be coprime to n. Show $\{a \times a_1, a \times a_2, \ldots, a \times a_{\phi(n)}\}$ is also a set of representatives of the classes coprime to n.*

Euler generalized Fermat's Little Theorem to give the following:

Theorem 2.17. *(Euler's Theorem) If a is relatively prime to n, then*

$$a^{\phi(n)} \equiv 1 \mod (n).$$

Proof. Let $\{a_1, a_2, \ldots, a_{\phi(n)}\}$ be a set of representatives of the classes coprime to n. Since a is coprime to n, Exercise 2.20 tells us

$$a_1 \times a_2 \times \cdots \times a_{\phi(n)} \equiv (a \times a_1) \times (a \times a_2) \times \cdots \times (a \times a_{\phi(n)}) \mod (n).$$

Simplifying:

$$a_1 \times a_2 \times \cdots \times a_{\phi(n)} \equiv a_1 \times a_2 \times \cdots \times a_{\phi(n)} \times a^{\phi(n)} \mod (n).$$

Notice $a_1, a_2, \ldots, a_{\phi(n)}$ are all coprime to n. Corollary 2.2 tells us that we can cancel these from both sides, giving

$$1 \equiv a^{\phi(n)} \mod (n).$$

□

2.11 FERMAT'S LAST THEOREM

One of the comments written in the margin of Fermat's *Arithmetica* became known as *Fermat's Last Theorem.* Though Fermat claimed to have a proof, it seems unlikely. It took over 350 years before Andrew Wiles finally gave a proof. During this time, Fermat's Last Theorem was probably the most famous unsolved mathematics problem.

Theorem 2.18. *The equation $x^n + y^n = z^n$ has no positive integer solutions if $n > 2$.*

Though there is no record of Fermat's 'marvelous proof', Fermat did prove the special case: there are no positive integer solutions when $n = 4$. First, he proved the following theorem.

Theorem 2.19. *There are no positive integer solutions to $x^4 + y^4 = z^2$.*

We will follow Fermat, using *infinite descent*. The idea is to show that if we have a solution, we can find another solution with a smaller positive z value. We can repeat the process with the new solution, obtaining another solution with a smaller value of z. We can keep doing this—the infinite descent. But there are only finitely many positive integer values smaller than the original z, so infinite descent is impossible.

Proof. Suppose there is a solution $a^4 + b^4 = c^2$ where a, b, and c are positive integers, then (a^2, b^2, c) is a Pythagorean triple. Euclid's formula (theorem 2.10) tells us there are coprime numbers m and n with

$$a^2 = m^2 - n^2, \; b^2 = 2mn, \text{ and } c = m^2 + n^2.$$

We can rearrange $a^2 = m^2 - n^2$ to obtain $a^2 + n^2 = m^2$, telling us that (a, n, m) is a Pythagorean triple. Applying Euler's formula once more gives coprime numbers m_1 and n_1 with

$$a = m_1^2 - n_1^2, \; n = 2m_1n_1, \text{ and } m = m_1^2 + n_1^2.$$

Plugging these expressions for m and n into $b^2 = 2mn$ gives

$$b^2 = 4m_1n_1(m_1^2 + n_1^2). \tag{2.14}$$

We know m_1, n_1, and $(m_1^2 + n_1^2)$ are pairwise coprime, from which we deduce that if a prime p divides b it will divide one and only one of m_1, n_1, or $(m_1^2 + n_1^2)$. Since the primes in b^2 must have even powers, we deduce each of m_1, n_1, and $(m_1^2 + n_1^2)$ must be an integer squared. Let

$$a_1^2 = m_1, \; b_1^2 = n_1, \text{ and } c_1^2 = m_1^2 + n_1^2.$$

We can substitute a_1 and b_1 into the formula for c_1^2 to obtain

$$a_1^4 + b_1^4 = c_1^2.$$

We started with the Pythagorean triple (a^2, b^2, c) and we ended with Pythagorean triple (a_1^2, b_1^2, c_1). Since

$$c = m^2 + n^2 > m^2 = (m_1^2 + n_1^2)^2 = c_1^4 \geq c_1,$$

we know that c_1 is smaller than c.

We can keep repeating the process, obtaining the Pythagorean triples (a_k^2, b_k^2, c_k) with $0 < c_k < c_{k-1}$ for all natural numbers k. But this is impossible, there are only finitely many positive integers less than c. Our initial assumption that there is a positive integer solution to $x^4 + y^4 = z^2$ is false. □

This result gives a proof of Fermat's Last Theorem for $n = 4$.

Corollary 2.3. *There are no positive integer solutions to* $x^4 + y^4 = z^4$.

Proof. If a, b, c satisfies $a^4 + b^4 = c^4$, then a, b, c^2 would be a solution of $x^4 + y^4 = z^2$. □

A similar argument gives:

Corollary 2.4. *There are no positive integer solutions to $x^{4k} + y^{4k} = z^{4k}$, for any positive integer k.*

Similarly, if there are no positive integer solutions to

$$x^m + y^m = z^m$$

for some m, then there are no positive integer solutions to

$$x^{km} + y^{km} = z^{km}$$

for any positive integer k. This means we only need to prove the theorem for cases where n is an odd prime.

As before, the fact that if (a, b, c) is a solution, then (ka, kb, kc) will also be a solution for any integer K, means we need only look for solutions a, b, and c that are pairwise coprime.

Euler proved there were no positive integer solutions for

$$x^3 + y^3 = z^3.$$

Sophie Germain

Sophie Germain (1776–1831) was a French mathematician who lived in Paris. She became interested in mathematics from reading books in her father's library. When she was 18, L'Ecole Polytechnique opened. This was a school for mathematics, physics, and chemistry, designed to produce engineers. Only men were allowed to attend the lectures. Germain requested lecture notes under the assumed name of Antoine-August LeBlanc and communicated with the professors by letter.

Joseph-Louis Lagrange (1736–1813) was one of the faculty. He quickly realized that LeBlanc was a talented mathematician and wanted to meet. Lagrange must have been surprised to find that LeBlanc was a pseudonym and Germain a woman, but he continued to work with her. This pattern would repeat. She first corresponded with Gauss under the name LeBlanc. When Germain eventually revealed who she was, Gauss was amazed, writing:

> ... how can I describe my astonishment and admiration on seeing my esteemed correspondent M. LeBlanc metamorphosed into this celebrated person ... when a woman, because of her sex, our customs and prejudices, encounters infinitely more obstacles than men in familiarizing herself with [number theory's] knotty problems, yet overcomes these fetters and penetrates that which is most

FIGURE 2.4 Sophie Germain

> hidden, she doubtless has the most noble courage, extraordinary talent and superior genius. ... The scientific notes with which your letters are so richly filled have given me a thousand pleasures. I have studied them with attention ...[3]

Germain worked on Fermat's Last Theorem. Instead of proving it for various given values of the exponents, she wanted to find a more general method. For a given p, if there is a solution (a, b, c) in positive integers to $x^p + y^p = z^p$, where a, b and c are pairwise coprime, there are two cases: case 1, p does not divide any of a, b, or c; or case 2, p divides exactly one of a, b, or c. She decided to concentrate on the first case.

Example 2.5. *We will show there are no solutions to $x^3 + y^3 = z^3$ for case 1. We are looking for solutions (a, b, c), where none of a, b, or c is divisible by 3. If there were a solution, each of these numbers must belong to one of the following congruence classes modulo 9:*

$$1, 2, 4, 5, 7, 8 \mod (9).$$

[3]Mackinnon, Nick (1990). "Sophie Germain, or, Was Gauss a feminist?". The Mathematical Gazette. 74 (470): 346-351.

Cubing these gives:

$$1^3 \equiv 4^3 \equiv 7^3 \equiv 1 \mod (9), \qquad 2^3 \equiv 5^3 \equiv 8^3 \equiv 8 \mod (9).$$

If we have a solution (a, b, c) and neither a nor b are divisible by 3, then a^3 is either 1 or 8 modulo 9. Similarly, b^3 is either 1 or 8, but then $a^3 + b^3$ is not going to equal either 1 or 8 modulo 9.

Germain proved the following theorem.

Theorem 2.20. *If p and $2p + 1$ are both odd primes, there are no solutions to $x^p + y^p = z^p$ for case 1.*

From a slightly more general version of this theorem, she proved there were no solutions to $x^p + y^p = z^p$ for any odd prime less than 100 for the first case.

If both p and $2p + 1$ are primes, we call p a *Sophie Germain prime.* The first few Sophie Germain primes are 2, 3, 5, 11, 23, and 29. It is conjectured that there are infinitely many, but nobody has proved this. Germain did not publish her theorem; it came to light after Adrien-Marie Legendre used it to prove Fermat's Last Theorem for $n = 5$ and credited her with the result in a footnote.

Lamé and Kummer

The French mathematician, Gabriel Lamé (1795–1870), devised an ingenious approach using complex roots of unity.[4] We define

$$\zeta_n = \cos\left(\frac{2\pi}{n}\right) + i \sin\left(\frac{2\pi}{n}\right)$$

then $\zeta_n^n = 1$, and we can write $x^n + 1$ as a linear product of factors:

$$x^n + 1 = (x + 1)(x + \zeta_n)(x + \zeta_n^2) \cdots (x + \zeta_n^{n-1}).$$

Given an odd prime p, Lamé extended the integers to form the *cyclotomic integers*, complex numbers of the form

$$a_0 + a_1\zeta_p + a_2\zeta_p^2 + \cdots + a_{p-1}\zeta_p^{p-1},$$

where $a_0, a_1, \ldots, a_{p-1}$ are integers.

Given a solution $a^p + b^p = c^p$, the left side now factors:

$$a^p + b^p = (a + b)(a + \zeta_p b)(a + \zeta_p^2 b) \cdots (a + \zeta_p^{p-1} b).$$

[4]The reader might want to consult Section 4.6 for background here.

In 1847, Lamé presented to the Paris Academy what he thought was a proof of Fermat's Last Theorem using cyclotomic integers. However, the proof assumed that, like integers, the cyclotomic integers had unique factorization. Joseph Liouville was in the audience and knew that Ernst Kummer was working on questions of factorization for extensions of the integers. For example, if we extend the integers by adding $\sqrt{-5}$ to form numbers of the form

$$a + b\sqrt{-5}, \qquad a, b \text{ are integers},$$

then we can factor 6 as $6 = 2 \times 3$ or as $6 = (1 + \sqrt{-5})(1 - \sqrt{-5})$.

Kummer knew that the cyclotomic integers did not always have a unique factorization. The first few odd primes work, but 23 gives the first example where unique factorization no longer holds.

This spurred Kummer's work on the problem. He defined the *class number* associated with the cyclotomic integers. This number gives a measure of how badly unique factorization fails. He defined *regular primes* to be those that do not divide their class number, and gave a proof of Fermat's Last Theorem for regular primes.

It is now known that there are infinitely many primes that are not regular, but whether there are infinitely many regular primes is still unknown.

The Proof

We have seen that work on Fermat's Last Theorem involved complex numbers and sophisticated algebraic properties. In the twentieth century, topological and geometric ideas came into play.

If we graph $x^n + y^n = 1$ in the plane, we get a curve. If we allow x and y to be complex numbers, then the graph is a surface. The *genus* of a surface counts the number of "holes" the surface has: a sphere has genus 0, a torus has genus 1, a torus with an added handle has genus 2, and so on. The genus of the surface given by $x^n + y^n = 1$ is

$$\frac{(n-1)(n-2)}{2}.$$

In particular, when $n \geq 4$, the genus is greater than 2.

1n 1922, Louis Mordell conjectured that if a polynomial $p(x, y)$ has rational coefficients, and the surface given by $p(x, y) = 0$ has genus greater than 2, then it only has finitely many rational solutions. This conjecture, if true, would tell us that $x^n + y^n = 1$ has only finitely many rational solutions when $n \geq 4$. This, in turn, would tell us that $x^n + y^n = z^n$ has only finitely many pairwise coprime integer solutions.

In 1983, Gerd Faltings proved Mordell's conjecture.

in 1957, Yutaka Taniyama and Goro Shimura conjectured that if polynomial of degree 3 with distinct roots (genus 1) has rational coefficients, then it must be *modular*. Being modular means that it can be constructed using special types of functions defined on the upper half of the complex plane.

Suppose we have an odd prime p and positive integers a, b, c with $a^p + b^p = c^p$, then it has a *Frey curve* defined by

$$y^2 = x(x - a^p)(y - b^p).$$

The curve is named after Gerhard Frey who first started studying it and pointing out its unusual properties. Frey thought these curves were not modular. He along with Jean-Pierre Serre and Ken Ribet eventually proved this was in fact the case.

Frey curves, if they exist, are polynomials of degree three with genus 1. The Taniyama–Shimura conjecture says that Frey curves must be modular. However, Frey, Serre, and Ribet showed that if Frey curves exist, they cannot be modular. Consequently, if the Taniyama–Shimura conjecture could be proved, Frey curves do not exist, and Fermat's Last Theorem would be proved.

In 1993, Andrew Wiles outlined a proof of the Taniyama–Shimura conjecture for the case of polynomials arising from Fermat's Last Theorem. It was finally finished with some help from Richard Taylor in 1995. This 200 page paper finally proved Fermat's Last Theorem.

Work continued in proving Taniyama–Shimura conjecture for the remaining cases. The work was finished in 1999. The Taniyama–Shimura conjecture is now known as the *modularity theorem*.

2.12 IRRATIONAL NUMBERS

Babylonian clay tablets from around 1800 BC list the lengths of the sides of right triangles. As we mentioned earlier, Plimpton 322 is a tablet with a list of Pythagorean triples. A natural question to ask is about the length of the hypotenuse of a triangle whose base and height were both equal to 1. The hypotenuse has length $\sqrt{2}$, so we do not have a Pythagorean triple. However, another clay tablet, YBC 729, from around 1800 to 1600 BC, gives an approximation to this length in sexagesimal notation. It is accurate to about six decimal places.

The Pythagoreans around 500 BC are credited with the discovery that $\sqrt{2}$ is not a rational number. Their motto was said to be "all is number." Numbers, at this time meant natural numbers or ratios of natural numbers. To have a proof that the length of the hypotenuse must have been a shock. There are stories about the discoverer being drowned.

The fact that $\sqrt{2}$ is not a rational number is easily proved.

Proof. We start by supposing $\sqrt{2}$ is rational and derive a contradiction, showing our initial supposition must be false.

If $\sqrt{2}$ is rational, then we can find integers m and n with $m/n = \sqrt{2}$. Moreover, since any fraction can be written in lowest terms, we can choose m and n *to have no common factors.*

If $m/n = \sqrt{2}$, then

$$m^2 = 2n^2, \tag{2.15}$$

so m^2 is even. But if m^2 is even, then so is m, which tells us we can find a positive integer k with $m = 2k$. Replacing m by $2k$ in equation 2.15 gives:

$$4k^2 = 2n^2.$$

Canceling 2 from both sides gives

$$2k^2 = n^2.$$

This tells us that n^2 is even from which we deduce that n must be even.

We now have our contradiction: m and n have no common factors, but they both have a factor of 2.

□

Exercise 2.21. *Show that $\sqrt{p}$ is irrational for every prime p.*

We can generalize this.

Theorem 2.21. *If r is a root of the equation $x^n + c_{n-1}x^{n-1} + c_{n-2}x^{n-2} + \cdots + c_0 = 0$ where all the coefficients are integers, then r is either an integer or it is irrational.*

Proof. Suppose that a/b is a root, where a and b are coprime integers.

$$\left(\frac{a}{b}\right)^n + c_{n-1}\left(\frac{a}{b}\right)^{n-1} + c_{n-2}\left(\frac{a}{b}\right)^{n-2} + \cdots + c_0 = 0.$$

Multiplying by b^n and rearranging gives:

$$a^n = -b\left(c_{n-1}a^{n-1} + c_{n-2}ba^{n-2} + \cdots + c_0b^{n-1}\right).$$

The right side is divisible by b, so a^n must be divisible by b. Since a and b are coprime, b is either 1 or -1, and a/b must be an integer.

□

If m is a positive integer, that is not a square, then there is no integer that will satisfy $x^2 - m = 0$. Theorem 2.21 tells us that $\sqrt{m}$ must be irrational.

Exercise 2.22. *If m is an integer that is not an nth power of another integer, show that $\sqrt[n]{m}$ is irrational.*

We began the chapter with a discussion of the Euclidean algorithm using repeated subtraction. Euclid describes the process in terms of lengths of lines in which the shorter line is deleted from the longer line. The algorithm appears in Book VII and Book X of the *Elements.* In Book VII, it is used to find the greatest common divisor of two numbers. Given a number, you can construct a line with that length. Book X is devoted to incommensurable lengths. In this book, Euclid shows if the two lengths are incommensurable, then the Euclidean algorithm does not halt. It is important to note that Book VII is about numbers, but Book X is about geometry. Euclid did not extend the idea of number to include irrational numbers, but rather took the existence of incommensurate lengths as a geometrical fact that showed not all lengths were numbers.

The Euclidean algorithm gives a way of expressing real numbers in terms of *continued fractions.* To illustrate, we return to the Euclidean algorithm on the pair $(114, 42)$ we looked at in example 2.2.

$$\begin{aligned} 114 &= 2 \times 42 + 30 \\ 42 &= 30 + 12 \\ 30 &= 2 \times 12 + 6 \\ 12 &= 2 \times 6 + 0. \end{aligned}$$

We use this calculation to obtain the continued fraction expansion for $114/42$. The first line tells us

$$\frac{114}{42} = 2 + \frac{30}{42}.$$

We can rewrite this as:

$$\frac{114}{42} = \frac{2}{1 + \frac{42}{30}}.$$

We use the second line of the calculation to obtain:

$$\frac{114}{42} = 2 + \frac{1}{1 + \frac{12}{30}}.$$

We can rewrite this as

$$\frac{114}{42} = 2 + \frac{1}{1 + \frac{1}{\frac{30}{12}}}.$$

The third line of the calculation gives:

$$\frac{114}{42} = 2 + \frac{1}{1 + \frac{1}{2 + \frac{6}{12}}}.$$

Finally, from the last line we obtain:

$$\frac{114}{42} = 2 + \cfrac{1}{1 + \cfrac{1}{2 + \cfrac{1}{2+0}}}.$$

A *simple continued fraction* has the form:

$$a_0 + \cfrac{1}{a_1 + \cfrac{1}{a_2 + \cfrac{1}{a_3 + \cfrac{1}{a_4 + \cdots}}}}$$

To save space we denote this as $[a_0; a_1, a_2, a_3, \ldots]$. So,

$$\frac{114}{42} = [2; 1, 2, 2].$$

To go from $[2; 1, 2, 2]$ to a fraction, you start with the right two numbers and then work to the left.

$$[2, 2] = 2 + \frac{1}{2} = \frac{5}{2}$$

$$[1, 2, 2] = 1 + \frac{2}{5} = \frac{7}{5}$$

$$[2; 1, 2, 2] = 2 + \frac{5}{7} = \frac{19}{7}.$$

You obtain the original rational in reduced form.

Finite continued fractions correspond to rational numbers—the Euclidean algorithm halts. Infinite continued fractions correspond to irrational numbers.

As an example, we will find the simple continued fraction for $\sqrt{2}$ which gives another proof that it is irrational.

$$\sqrt{2} = 1 + (\sqrt{2} - 1) = 1 + \frac{1}{1 + \sqrt{2}}.$$

Replacing the $\sqrt{2}$ in the denominator on the right by the expression for $\sqrt{2}$ gives

$$\sqrt{2} = 1 + (\sqrt{2} - 1) = 1 + \frac{1}{1 + 1 + \frac{1}{1+\sqrt{2}}} = 1 + \frac{1}{2 + \frac{1}{1+\sqrt{2}}}.$$

Repeating, we obtain

$$\sqrt{2} = 1 + \cfrac{1}{2 + \cfrac{1}{2 + \cfrac{1}{2 + \cfrac{1}{2 + \cdots}}}}$$

So, $\sqrt{2} = [1; 2, 2, 2, \ldots]$. Since, this is infinite, $\sqrt{2}$ is irrational.

Later in the book we will talk about the number e and why it is important. Euler showed that e has continued fraction

$$[2; 1, 2, 1, 1, 4, 1, 1, 6, 1, 1, 8, \ldots],$$

where you keep adding three terms: the first two are 1s, the third is two more than the previous third term. This gives an infinite continued fraction and is how Euler first proved that e is irrational in 1737.

The first proof that π is irrational also used continued fractions. However, the continued fraction expansion for π is $[3; 7, 15, 1, 292, \ldots]$. There is no apparent pattern, so instead of using π Johann Heinrich Lambert turned to the function $\tan(x)$. In 1761, he proved

$$\tan(x) = \cfrac{x}{1 - \cfrac{x^2}{3 - \cfrac{x^2}{5 - \cfrac{x^2}{7 - \cdots}}}}.$$

He used this and $\tan(\pi/4) = 1$ to show π is irrational.

Continued fractions have other nice properties, one is they give best approximations. We say that a rational number p/q is a *best approximation* to a real number r, if the distance between r and p/q is less than any other rational number whose denominator is less than or equal to q.

Given a continued fraction $[a_0; a_1, a_2, \ldots]$, we call the rational number $[a_0; a_1, a_2, \ldots, a_m]$, the mth *convergent*. The following theorem, which we won't prove, states that the mth convergents are best approximations.

Theorem 2.22. *Given a real number r with continued fraction expansion $[a_0; a_1, a_2, \ldots]$, let p_m/q_m denote its mth convergent, then p_m/q_m is the best rational approximation to r with denominator less than or equal to q_m.*

We look at $\pi = 3.141592654\ldots$. Its continued fraction expansion is $[3; 7, 15, 1, 292, \ldots]$. From this we obtain:

$$m_1 = [3; 7] = \frac{22}{7} = 3.142857143\ldots$$

$$m_2 = [3; 7, 15] = \frac{333}{106} = 3.141509434\ldots$$

$$m_3 = [3; 7, 15, 1] = \frac{355}{113} = 3.14159292\ldots$$

This means if we restrict to denominators of 7 or less, then 22/7 is the best approximation to π. If we restrict to denominators of 106 or less, then 333/106 is the best approximation to π; and so on.

Algebraic numbers and transcendental numbers

We say r is an algebraic number if it is the root of a polynomial with integer coefficients. Looking at linear polynomials of the form $nx - m$ shows all rational numbers are algebraic. Given any positive integers m and n, the polynomial $x^n - m$ tells us that $\sqrt[n]{m}$ is algebraic. The equation $ax^2 + bx + c$ shows

$$\frac{-b \pm \sqrt{b^2 - 4ac}}{2a}$$

is algebraic.

The *degree* of an algebraic number α is the degree of the smallest degree polynomial for which α is a root. So, for example, $\sqrt{3}$ has degree 2.

Real numbers that are not algebraic are called *transcendental*. When Lambert proved π was irrational, he conjectured it was transcendental, but could not prove it. At that time, nobody had proved transcendental numbers existed. There were no examples. Joseph Liouville constructed the first known transcendental number in 1844.

Liouville considered approximations to irrational numbers by rational numbers. He showed that if an irrational number was algebraic then rational approximations would have to satisfy the following condition:

Theorem 2.23. *(Liouville) Let α be an algebraic irrational number with degree d. Then we can find $c > 0$ such that*

$$\left|\alpha - \frac{p}{q}\right| \geq \frac{c}{q^d}$$

for any rational number $\frac{p}{q}$.

Liouville had shown that there was a limit on how well an algebraic rational number could be approximated by rationals. He then constructed an irrational number in such a way that it can be approximated extremely closely by rationals. He defined L, by

$$L = \sum_{k=1}^{\infty} \frac{1}{10^{k!}}.$$

It has 1s in its decimal expansion at the $k!$ places and 0s elsewhere. Recall

$$1! = 1, \quad 2! = 2 \times 1 = 2, \quad 3! = 3 \times 2 \times 1 = 6, \quad 4! = 4 \times 3 \times 2 \times 1 = 24,$$

$$\text{so } L = 0.110001000000000000000001 00\ldots.$$

The decimal expansion consists of 1s separated by ever larger strings of 0s.

Theorem 2.24. *(Liouville) L is a transcendental number.*

Proof. Let

$$L_m = \sum_{k=1}^{m} \frac{1}{10^{k!}}.$$

We will denote it by $\frac{p}{q}$, where $q = 10^{m!}$. Now,

$$\left| L - \frac{p}{q} \right| \leq \frac{1}{10^{(m+1)!}} = \frac{1}{q^{m+1}}.$$

If L were algebraic, by Theorem 2.23, we would have

$$\left| L - \frac{p}{q} \right| \geq \frac{c}{q^d}$$

for some fixed $c > 0$. However, for any $c > 0$, we can choose m large enough that

$$\frac{1}{q^{m+1}} < \frac{c}{q^d},$$

telling us that L cannot be algebraic and so must be transcendental.

□

In showing L is transcendental, Liouville proved the existence of transcendental numbers. He did this in 1844. In 1873, Charles Hermite proved e is transcendental, and in 1882, Ferdinand von Lindeman gave the first proof that π is transcendental. The most remarkable result from this time is due to Georg Cantor who showed that transcendental numbers are far from mathematical oddities but are ubiquitous. We will look at Cantor's work in a later chapter.

Suggestion for Further Reading

Carl B. Boyer, Uta C. Merzbach. *A History of Mathematics.* 3rd Edition. Wiley, 2011.

John Stillwell. *Mathematics and its History* Undergraduate Texts in Mathematics. 3rd Edition. Springer, 2010.

CHAPTER 3

Medieval and Renaissance Mathematics

In this chapter, we describe some of the major developments in mathematics in the period following the foundational contributions of the ancient world and prior to the introduction of calculus.

This period saw important contributions from many parts of the world in the areas of number theory, algebra, and geometry. Highlights include the work of Indian and Chinese mathematicians in solving Diophantine equations (equations with integer coefficients and solutions), contributions of Fibonacci, and the invention of logarithms.

Of crucial importance for the further development of mathematics was the introduction of a positional number system and the invention of symbolic algebra (polynomials with definite but unspecified coefficients, general formulas for the solutions, etc.). Also, Descartes' invention of analytic geometry: the system of defining curves by equations, with reference to co-ordinates in the (*Cartesian*) plane and methods for the calculations of areas under graphs and slopes of tangent lines devised by Fermat, Cavalieri, Barrow, and Wallis, and their contemporaries. Their work extended the method of Archimedes and paved the way for calculus. These developments and more are discussed in this chapter.

3.1 MATHEMATICS OF THE MID-EAST

We have discussed some of the contributions of al-Khwarizmi and Omar Khayyam in earlier chapters. The purpose of this section is to present the

DOI: 10.1201/9781003470915-3

work of two other great Muslim mathematicians, Abû Kâmil and Thâbit ibn Qurra.

Abû Kâmil (circa 850–930) was of Egyptian descent and worked in the period following al-Khwārizmī. His book *Kitâb fil-jabra w'al muqâbala (Book of Algebra)* is a commentary on, and a continuation of al-Khwarizmi's work. The book contains a total of 69 problems in algebra. Kâmil elaborated on many of the problems that al-Khwārizmī had discussed and added new methods of his own. One of the problems addressed by Kâmil in the *Algebra* is to solve simultaneously the two equations

$$x + y = 10,$$
$$\frac{x}{y} + \frac{y}{x} = \frac{17}{4}.$$

Kâmil converts the second equation into

$$x^2 + y^2 = \frac{17xy}{4},$$

then substitutes $y = 10 - x$ to obtain the quadratic equation

$$\frac{25}{4}x^2 + 100 = \frac{125}{2}x.$$

Kâmil then introduces a new method to solve this equation: set $x = 5 - z$. (The point of this substitution is to make the term in z disappear from the equation.) Under this substitution, the equation becomes

$$50 + 2z^2 = \frac{17}{4}(25 - z^2).$$

Solving this equation gives $z = 3$. Hence $x = 2$ and $y = 8$.

Kâmil also developed a new method for combining square roots. A major advance in his work is the formulation of equations with irrational coefficients. An example is a problem in the *Algebra* in which he asks for a number, such that, if the square root of 3 is added to it and the square root of 2 is added to it, and the two numbers are multiplied, then the result will be 20. Thus, a solution of the equation

$$(x + \sqrt{3})(x + \sqrt{2}) = 20.$$

Kâmil gives the correct solution to the equation,

$$x = \sqrt{\frac{85}{4} - \sqrt{6} + \sqrt{\frac{3}{2}}} - \sqrt{\frac{3}{4}} - \sqrt{\frac{1}{2}}.$$

Another distinguished Arabic scholar from this era was Thâbit ibn Qurra (circa 836–901). Thâbit made contributions to both number theory and geometry. Thâbit's treatise, *Book on the Determination of Amicable Numbers,* contains 10 propositions. A pair of amicable numbers are two numbers, each of which is equal to the sum of the proper divisors of the other. One of Thâbit's propositions is the following rule for constructing pairs of amicable numbers: *if* $p = 3 \cdot 2^n - 1, q = 3 \cdot 2^{n-1} - 1,$ *and* $r = 9 \cdot 2^{2n-1} - 1$ *are all prime, then* $M = 2^n pq$ *and* $N = 2^n r$ *form a pair of amicable numbers.* An example is $n = 2$. In this case $p = 3 \cdot 2^2 - 1 = 11, q = 3 \cdot 2 - 1 = 5$, and $r = 9 \cdot 2^3 - 71$ are all prime, thus the pair $M = 2^2 \cdot 11 \cdot 5 = 220$ and $N = 2^2 \cdot 71 = 284$ are amicable. And, indeed, the proper divisors of 220: 1, 2, 4, 5, 10, 11, 20, 22, 44, 55, 110 sum to 284, while the proper divisors of 284: 1, 2, 4, 71, 142 sum to 220.

Thâbit gave the following generalization of the Pythagorean theorem, from right-angled to arbitrary triangles.

Theorem 3.1. *Suppose that in figure 3.1 the angles* $\angle BAC, \angle AC'B', \angle AB'C'$ *are equal. Then*

$$AB^2 + AC^2 = (BC)(BB' + CC').$$

Proof. Denote $\angle BAC = \angle AC'B' = \angle AB'C' = \theta$. It follows from the law of cosines that

$$\begin{aligned} AB^2 + AC^2 &= BC^2 + 2(AB)(AC)(\cos\theta) \\ &= BC^2 + (AB)(AC)(\cos\theta + \cos\theta) \\ &= BC^2 + (AB)(AC)\Big(\frac{DC'}{AC'} + \frac{DB'}{AB'}\Big) \\ &= BC^2 + (AB)(AC)\Big(\frac{DC' + DB'}{AB'}\Big). \end{aligned}$$

From the similarity of $\triangle ABC$ and $\triangle AB'B$, we have $AB/AB' = BC/AC$, hence

$$\begin{aligned} AB^2 + AC^2 &= BC^2 + (BC)(DC' + DB') \\ &= (BC)(BC + DC' + DB') \\ &= BC(BC' + C'D + DB' + B'C + DC' + DB') \\ &= (BC)(BB' + CC') \end{aligned}$$

as claimed. □

In a work, entitled *Book on the Measurement of the Conic Section Called Parabolic*, Thâbit proved that the area of a parabolic segment is two thirds the product of its base and its height. He did this 8y inscribing triangles whose bases were proportional to the sum of the odd integers and used the summation formula

$$1 + 3 + 5 + 7 + \cdots + (2n - 1) = n^2. \tag{3.1}$$

Exercise 3.1. *Deduce the Pythagorean theorem as a corollary of Thâbit's theorem.*

Yet another work of Thâbit's, *The Proof of the Well-known Postulate of Euclid* is particularly interesting from a more modern perspective. Here Thâbit attempted to prove Euclid's fifth postulate from the other four postulates. In doing so, he introduced, for the first time, quadrilaterals in which the two base angles are right angles and the two vertical sides have equal length. These are the rectangles now known as *Saccheri rectangles* after the work of the eighteenth century Italian geometer Giovanni Saccheri (who attempted a similar task), which later came to play a major role in the development of non-Euclidean geometry.

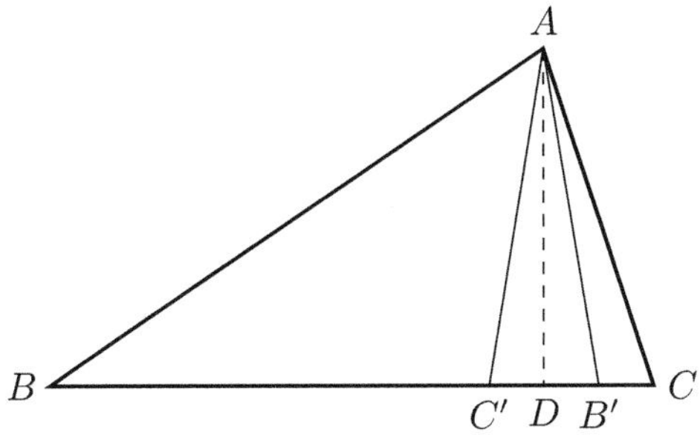

FIGURE 3.1 Thâbit's theorem

3.2 FIBONACCI

Leonardo of Pisa, who went by the name of Fibonacci (a contraction of *filius Bonaccio*, "son of Bonaccio") is considered the greatest mathematician of the Middle Ages. Fibonacci was born in Pisa about 1175 and educated in North Africa, where his father was in charge of a customhouse. As a young man he traveled widely in the countries of the Mediterranean, where he encountered the various arithmetical systems used in the commerce of the different countries. Fibonacci would have soon recognized the enormous technical advantages of the Hindu-Arabic decimal systems, with its positional notation and symbol for zero, over the clumsy Roman system in use at that time in Italy. On his return to Pisa in 1202, Fibonacci wrote his famous treatise *Liber Abaci* (Book of Counting) and it is largely through the second edition of this work, which appeared in 1228, that the non-Muslim world became acquainted with Arabic numerals.

Fibonacci later compiled another notable work, the *Liber Quadratorum* (Book of Squares), in which he studies diophantine equations of the second degree. Fibonacci had been presented at the court of Emperor Frederick II

FIGURE 3.2 Leonardo of Pisa—Fibonacci (*courtesy of Columbia University, David Eugene Smith Collection*)

and one of Frederick's retinue had assigned him a problem as a test of his mathematical skill: to find a number, which, increasing or decreasing its square by 5, would again result in a square. Fibonacci gave the correct answer, $\frac{41}{12}$,

$$\left(\frac{41}{12}\right)^2 + 5 = \left(\frac{49}{12}\right)^2, \quad \left(\frac{41}{12}\right)^2 - 5 = \left(\frac{31}{12}\right)^2.$$

Problems of this type would no doubt have stimulated Fibonacci's interest in the subject matter of the *Liber Quadratorum*, published in 1225. In fact, an example of a problem treated in *Liber Quadratorum* is the following, closely related problem: solve, in rational numbers, the pair of equations

$$x^2 + x = u^2,$$
$$x^2 - x = v^2.$$

Fibonacci solves the problem by first finding three squares a^2, b^2, c^2 with a common difference d, i.e.,

$$a^2 = b^2 - d, \quad c^2 = b^2 + d,$$

then setting $x = b^2/d$. This results in

$$x^2 + x = \frac{b^4}{d^2} + \frac{b^2}{d} = \frac{b^2(b^2 + d)}{d^2} = \frac{b^2c^2}{d^2} = \left(\frac{bc}{d}\right)^2,$$
$$x^2 - x = \frac{b^4}{d^2} - \frac{b^2}{d} = \frac{b^2(b^2 - d)}{d^2} = \frac{b^2a^2}{d^2} = \left(\frac{ba}{d}\right)^2.$$

For example, taking $a^2 = 1, b^2 = 25, c^2 = 49$, results in the solution $x = 25/24$ to the equations

$$x^2 + x = \left(\frac{35}{24}\right)^2,$$
$$x^2 - x = \left(\frac{5}{24}\right)^2.$$

In a publication of 1224, entitled *Flos* ("flower"), Fibonacci studied 15 diophantine equations, including the cubic equation

$$x^3 + 2x^2 + 10x = 20,$$

which he treats in the form

$$x + \frac{x^2}{5} + \frac{x^3}{10} = 2$$

and shows, by a typically ingenious argument, that this equation has no rational solutions.

Fibonacci's argument is by contradiction and goes as follows. Suppose there is a rational solution to the equation, and express it in the form $x = a/b$ where $\gcd(a, b) = 1$. (Since it is clear the equation has no integer solutions, $b \neq 1$). Then the expression

$$\frac{a}{b} + \frac{a^2}{5b^2} + \frac{a^3}{10b^3} = \frac{a(10b^2 + 2ab + a^2)}{10b^3}$$

is an integer. Since a and b are co-prime, this implies that b^3, and hence b divides $10b^2 + 2ab + a^2$, which in turn implies that b divides a^2. Using prime factorization, it is easy to see that this implies b divides a, contradicting $\gcd(a, b) = 1$.

Fibonacci went further in investigating the nature of the irrational roots of the equation. By a case by case ruling out of certain Euclidean irrationals, such as $a + \sqrt{b}$, $\sqrt{\sqrt{a} + \sqrt{b}}$,... Fibonacci convinced himself, and stated, that the roots could not be constructed using only straightedge and compass. While a full proof of results such as this would need to wait hundreds of years until the development of field theory (as we will see in Chapter 4), the observation is interesting as the first intimation that there exist numbers that transcend the classical Euclidean geometric framework. Fibonacci went on to obtain the remarkably accurate approximation to the root, $x = 1.33688081075...$, accurate to nine decimal places.

While Fibonacci never formally accepted the notion of negative numbers in mathematics, he did make a step in this direction in *Flos*, interpreting them in financial terms as debits as opposed to credits.

It is ironic that, despite Fibonacci's many and influential contributions to mathematics, his name is known today largely for a sequence attributed to him in the nineteenth century by the French mathematician Edouard Lucas. The sequence was mentioned only briefly in Fibonacci's works, in connection with a trivial problem involving rabbit reproduction.

The Fibonacci Sequence

The following problem is stated in *Liber Abaci*: *A man put one pair of rabbits in a certain place entirely surrounded by a wall. How many pairs of rabbits can be produced by that pair in a year, if the nature of these rabbits is such that every month, each pair bears a new pair which from the second month on becomes productive?*

The solution is as follows, assuming that the original pair of rabbits is an infant pair and that none of the rabbits die. Nothing happens in the second month since the original pair of rabbits is only one month old and cannot reproduce. In the third month, the original pair of rabbits becomes mature and gives birth to a new pair. There are now two pairs of rabbits, a mature pair and a young pair. Next month, there are three mature pairs and the two pairs that were mature in the previous generation give birth to two new young pairs, etc. The number of rabbit pairs in the first year is shown in the following table.

Month n	**Pairs** F_n
1	1
2	1
3	2
4	3
5	5
6	8
7	13
8	21
9	34
10	55
11	89
12	144

It is clear that the number of new births in each nth month (>2) are the number of pairs present in month $n-2$ (all of which are mature by month n). Thus there is the following recursive rule for the formation of the sequence F_n,

$$F_n = F_{n-1} + F_{n-2}, \quad n > 2. \tag{3.2}$$

The entire sequence can be generated from this relation, together with the "initial conditions"

$$F_1 = 1, \quad F_2 = 1. \tag{3.3}$$

Exercise 3.2*. *Show that the "generating function" for the Fibonacci sequence*

$$F_1 + F_2 x + F_3 x^2 + \cdots = \frac{1}{1 - x - x^2}.$$

The Fibonacci sequence $\{F_n\}$ has some remarkable properties. We note and prove a few of them.

Theorem 3.2. *Any two consecutive terms of the sequence are relatively prime.*

Proof. Suppose there is an integer d that divides both F_n and $F_{n-1}, n \geq 3$. Then d divides the difference $F_n - F_{n-1} = F_{n-2}$. Thus d divides both F_{n-1} and F_{n-2}. Iterating this argument (arguing by "infinite descent", if you will), we see that d divides both F_2 and F_1. Hence $d = 1$. □

In fact, computing the ratio of two consecutive terms F_n/F_{n-1} of the Fibonacci sequence for large n reveals an interesting pattern. For example, $F_9/F_8 = 34/21 = 1.61904..., F_{12}/F_{11} = 1.6179..., F_{14}/F_{13} = 377/233 = 1.618025, ...$ As n becomes larger and larger, these ratios seem to get closer and closer to the golden ratio $\phi = (1 + \sqrt{5})/2 = 1.618033....$ In the language of *limits*, we would say

$$\lim_{n\to\infty} \frac{F_n}{F_{n-1}} = \phi.$$

This is confirmed by the following observation: Dividing (3.2) through by F_{n-1} gives

$$\frac{F_n}{F_{n-1}} = 1 + \frac{F_{n-2}}{F_{n-1}}. \tag{3.4}$$

Now assuming that $L = \lim_{n\to\infty} F_n/F_{n-1}$ exists, we will have $\lim_{n\to\infty} F_{n-2}/F_{n-1} = 1/L$. So taking the limit as n tends to infinity in equation 3.4 gives

$$L = 1 + \frac{1}{L},$$

whence we see that $L = \phi$. Thus the Fibonacci sequence can be used to produce rational approximations to the golden ratio, to within any prescribed degree of accuracy.

Another property, which was proved by Lucas, is

Theorem 3.3.

$$F_1 + F_2 + \cdots + F_n = F_{n+2} - 1.$$

For example, referring to the previous table, we have

$$F_1 + F_2 + \ldots F_7 = 1 + 1 + 2 + 3 + 5 + 8 + 13 = 33 = F_9 - 1.$$

The proof of Theorem 3.3 follows easily from the relations

$$\begin{aligned} F_1 &= F_3 - F_2 \\ F_2 &= F_4 - F_3 \\ F_3 &= F_5 - F_4 \\ &\vdots \\ F_{n-1} &= F_{n+1} - F_n \\ F_n &= F_{n+2} - F_{n+1}. \end{aligned}$$

Adding these equations, and noting the cancellations on the right-hand side (telescoping), we get the result.

Theorem 3.4.

$$F_n^2 = F_{n-1}F_{n+1} + (-1)^n, \quad n \geq 2.$$

For example, taking $n = 6$ and $n = 7$, we have

$$\begin{aligned} F_6^2 &= 8^2 = 5 \cdot 13 - 1 = F_5F_7 - 1, \\ F_7^2 &= 13^2 = 8 \cdot 21 + 1 = F_6F_8 + 1. \end{aligned}$$

Proof. We prove Theorem 3.4 as follows.

$$\begin{aligned} F_n^2 - F_{n-1}F_{n+1} &= F_n(F_{n-1} + F_{n-2}) - F_{n-1}F_{n+1} \\ &= (F_n - F_{n+1})F_{n-1} + F_nF_{n-2} \\ &= -F_{n-1}^2 + F_nF_{n-2}. \end{aligned}$$

Thus

$$F_n^2 - F_{n-1}F_{n+1} = (-1)(F_{n-1}^2 - F_{n-2}F_n).$$

Note that the expression inside the bracket in the right-hand side of the last equation is simply the left-hand side with n replaced by $n-1$. Iterating this relation, we eventually obtain, after $n-2$ steps, the relation,

$$\begin{aligned} F_n^2 - F_{n-1}F_{n+1} &= (-1)^{n-2}(F_2^2 - F_3F_1) \\ &= (-1)^{n-2}(1^2 - 2 \cdot 1) \\ &= (-1)^{n-2}(-1) \\ &= (-1)^{n-1}, \end{aligned}$$

which is what we sought to prove. □

Perhaps the most remarkable property of the Fibonacci sequence is the following closed-form for the terms, which goes under the name of *Binet's formula.*

Theorem 3.5.

$$F_n = \frac{1}{\sqrt{5}}\left(\frac{1+\sqrt{5}}{2}\right)^{n-1} - \frac{1}{\sqrt{5}}\left(\frac{1-\sqrt{5}}{2}\right)^{n-1}. \tag{3.5}$$

Proof. One way to prove Theorem 3.5 is to make use of the observations:

1. Solutions to the defining relation (3.2) can be obtained in the form $F_n = r^n$.

2. The relation is *linear*, i.e., if the sequences x_n and y_n both satisfy (3.2), then so does the sequence $ax_n + by_n$, for any constants a and b.

Substituting $F_n = r^n$ into (3.2) and reducing leads to the following quadratic equation for r

$$r^2 = r + 1.$$

Interestingly enough, this is precisely the equation we solved to find the golden ratio in Section 1.3. The solutions are

$$r_1, r_2 = \frac{1 \pm \sqrt{5}}{2}.$$

Thus every sequence of the form

$$F_n = ar_1^n + br_2^n$$

satisfies (3.2). We now choose a and b to satisfy the initial conditions (3.3). This results in the system of linear equations

$$a + b = 1,$$
$$ar_1 + br_2 = 1.$$

Solving for a and b yields Binet's formula (3.5). □

Exercise 3.3.[*] *Use (3.5) to show that*

$$\lim_{n\to\infty} \frac{F_{n+1}}{F_n} = \frac{1+\sqrt{5}}{2}.$$

Exercise 3.4.[*] *Use the foregoing method to obtain a closed form for the Lucas numbers L_n, defined by*

$$L_1 = 2, \quad L_2 = 1,$$

$$L_n = L_{n-1} + L_{n-2}, \quad n > 2.$$

The Fibonacci Sequence in Nature

A fascinating aspect of the Fibonacci sequence is its seemingly ubiquitous appearance in nature, in the structure of plants and in the shapes of certain creatures. For instance, many flowers, fruits, and vegetables exhibit patters of interlocking spirals running in both a clockwise and a counterclockwise direction. Examples are asparagus, pine cones, and sunflowers (see figures 3.3, 3.4, and 3.5). The numbers of clockwise and counterclockwise spirals almost invariably correspond to two consecutive numbers in the Fibonacci sequence. For example, there are 3 clockwise and 5 counterclockwise spirals on a typical asparagus tip. The pine cone generally has either 5 clockwise and 8 counterclockwise spirals, or 8 clockwise and 13 counterclockwise spirals, while a sunflower will most likely exhibit a pattern of 34 clockwise and 55 counterclockwise spirals at its center (also 55 and 89, or 89 and 144, depending on the size of the sunflower).

FIGURE 3.3 Asparagus tip

FIGURE 3.4 Pine cone

FIGURE 3.5 Sunflower

3.3 MATHEMATICS IN CHINA

The thirteenth century is regarded as the high point of traditional Chinese mathematics. However, China has a rich tradition of mathematics dating back to ancient times.

The most influential work in early Chinese mathematics is the *Nine Chapters of the Mathematical Art.* First assembled as a book at about the same time as Euclid's *Elements*, the book serves roughly the same purpose, as a compendium of the mathematics known to the Chinese at that time. The book survives today in the form of a commentary drawn up by Liu Hui in A.D. 263, in which he summarizes and expands on the content of the *Nine Chapters* and enriches it with new ideas of his own. The book became one of the earliest printed textbooks in 1084, when a printed version appeared using a wood block technique where each page was separately carved from one wooden block. The printed version of the book achieved wide circulation throughout China.

The *Nine Chapters* was never intended as a theoretical work in mathematics in the style of the Greeks, but rather as a practical manual giving guidance on problems that the ruling officials of the day were likely to encounter, in surveying, the construction of dikes and waterways, matters of commerce, taxation, etc. The early parts of the *Nine Chapters* gives rules (some incorrect) for computing the areas of familiar geometric shapes, rectangles, triangles, trapezoids, sectors of circles, and volumes of spheres, cylinders, pyramids, and circular cones. For example, the area of the circle is given as $\frac{3}{4}d^2$, where d is the diameter, which would only be correct if the value of π were taken as 3. There is the correct formula for the volume of a frustrum of a pyramid, also known to the Egyptians.

Problem 13 of Chapter 9 ("Right Angles") solves the following problem, which shows that the Chinese were in possession of the Pythagorean theorem.

There is a bamboo of 10 ch'ih high. It is broken and the upper end touches the ground 3 ch'ih away from the root. Find the height of the break.

In his commentary, Liu Hui obtains approximations to π by first inscribing a regular hexagon inside a circle, then successively doubling the number of sides, in the spirit of the method used by Archimedes. In this fashion, he shows that the area of the regular 192-gon inscribed in a circle of radius 10, is 314.1024. This determines a value for π of 3.141042. Two hundred years later, the mathematician-astronomer Tsu Chung-chi (430–501), in collaboration with his son, would use Liu Hui's method to estimate π as lying in the range

$$3.1415926 < \pi < 3.1415927.$$

This level of accuracy was not reached in the West until the end of the sixteenth century.

The eighth of the *Nine Chapters*, titled "The Way of Calculating by Arrays" provides the first systematic method for the solution of a set of simultaneous linear equations. An example is the first problem in the chapter, which reads as follows. *There are three grades of corn. After threshing, three bundles of top grade, two bundles of medium grade, and one bundle of low grade make 39 dou. Two bundles of top grade, three bundles of medium grade, and one bundle of low grade will produce 34 dou. The yield of one bundle of top grade, two bundles of medium grade, and three bundles of low grade is 26 dou. How many dou are contained in each bundle of each grade?*. The problem is equivalent to the system of linear equations

$$\begin{aligned} 3x + 2y + z &= 39, \\ 2x + 3y + z &= 34, \\ x + 2y + 3z &= 26. \end{aligned}$$

These equations were expressed, using rods on an a counting board, as the array

$$\begin{array}{ccc} 1 & 2 & 3 \\ 2 & 3 & 2 \\ 3 & 1 & 1 \\ 26 & 34 & 39 \end{array}$$

By performing appropriate multiplications and subtractions, this array was converted into the reduced form

$$\begin{array}{ccc} 0 & 0 & 3 \\ 0 & 5 & 2 \\ 36 & 1 & 1 \\ 99 & 24 & 39 \end{array}$$

The last array translates back into the equations

$$36z = 99,$$
$$5y + z = 24,$$
$$3x + 2y + z = 39,$$

whence it is determined that $z = 11/4, y = 17/4$, and $x = 37/4$.

This approach prefigures the method of *Gaussian elimination* introduced in the nineteenth century to solve systems of linear equations.

Exercise 3.5* *Solve the system of equations.*

$$x + y + z = 8,$$
$$2x + 3y + 4z = 27,$$
$$4x - 3y + 5z = 15.$$

Exercise 3.6* *Show that the system of equations*

$$ax + by = p,$$
$$cx + dy = q.$$

has either a unique solution, infinitely many solutions, or no solutions and give a condition which ensures that there is a unique solution.

Liu Hui also produced a treatise on surveying, titled *Sea Island Mathematical Manual.* This contained nine problems, typical of which is the first. *There is a sea island to be measured. Two poles each 30 feet high are erected on the same level, 1000 paces apart [1 pace = 6 feet], so that the rear pole is in a straight line with the island and the first pole. If a man walks 123 paces back from the first pole, the highest point of the island is just visible through the top of the pole when he views it from ground level. Should he move 127 paces back from the rear pole, the summit of the island is just visible through the top of the pole when seen from ground level. It is required to find the height of the island and its distance from the nearer pole.*

The problem is illustrated in figure 3.6. The dashed line FI is constructed to be parallel to the line BE. The given distances are $|AH| = |DG| = |EF| = 5$ (paces), $|DJ| = 123, |EC| = 127$. Thus $|EI| = 123$ and $|IC| = 4$.

We are required to find the height $|AB|$ and the horizontal distance $|AD|$. Denote these by x and y respectively. Observing that $\triangle BHG$ and $\triangle FEI$ are similar, as are $\triangle BGF$ and $\triangle FIC$, we have

$$\frac{x-5}{5} = \frac{y}{123} = \frac{1,000}{4},$$

from which we deduce that $x = 1{,}255$ paces and $y = 30{,}750$ paces.

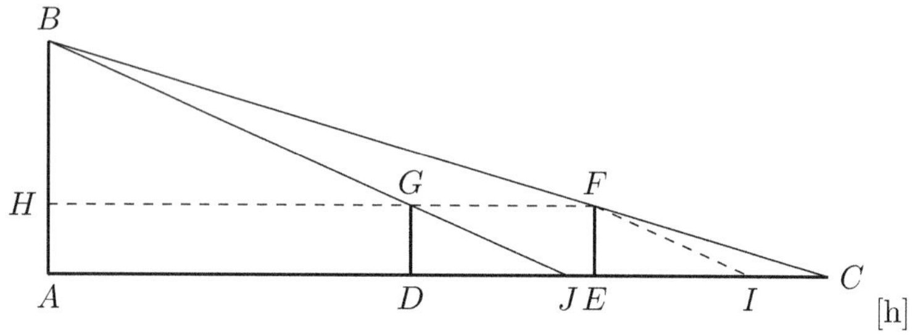

FIGURE 3.6 Liu's surveying problem

One of the leading Chinese mathematicians of the thirteenth century was Ch'in Chu-shao (1202–261) who published a celebrated work *Mathematical Treatise in Nine Sections* in 1247 (the Chinese evidently had a penchant for the number nine in their titles.) The *Nine Sections* is the oldest existing mathematical text to contain a round symbol for the number zero, and also the first to contain polynomial equations with degree larger than 3. For example, one of the problems discussed leads to the equation

$$x^{10} + 15x^8 + 72x^6 - 864x^4 - 11,664x^2 - 34,992 = 0,$$

for which Ch'in found the solution $x = 9$.

Another notable mathematician from this period is Li Ye (1192–1279). His book *Old Mathematics in Expanded Sections* contains 64 problems involving quadratic equations arising from geometric settings. An example is: *There is a circular pond in the middle of a square field, and the area outside the pond is 64 square pu. It is known only that the sums of the perimeters of the square and the circle is 300 pu. Find the perimeters of the square and the circle.*

Assuming the ancient value of 3 for π, Li solves the problem, obtaining the diameter of the pond as 20 pu.

Exercise 3.7. *Set up and solve a quadratic equation and thereby derive Li's solution to the problem.*

The last of the Chinese mathematicians that we discuss here is Chu Shih-chieh, who published two major treatises *Introduction to Mathematical Studies* in 1299 and the *Precious Mirror of the Four Elements* in 1303.

The *Precious Mirror* opens with a diagram showing the coefficients of the binomial expansions of $(1 + x)^n$ for values of n up to 8, essentially *Pascal's triangle.* In this work Chu treats polynomial equations of high degree, introducing a new method for approximating the solutions now known in the West as *Horner's method,* after the mathematician William Horner who rediscovered it in 1819. Chu uses as an example the quadratic equation

$$x^2 + 252x - 5292 = 0.$$

Chu finds by inspection that there is a root between 19 and 20. He then shifts the root by the substitution $y = x - 19$. The equation in y,

$$y^2 + 290y - 143 = 0$$

thus has a root between 0 and 1. Chu simplifies the equation by approximating the y^2 term by y and solving the resulting linear equation to obtain $y = 143/291$. Thus he obtains as an approximation to the original root, $x = 19 + 143/291$.

In the *Precious Mirror*, Chu also studies arrangements of balls in various geometric configurations, triangle, pyramids, cones, etc. and in the process, both anticipates the modern mathematical topic of "sphere packing," and gives the summation formulas

$$\begin{aligned}
&1 + 2 + 3 + \cdots + n = \frac{n(n+1)}{2},\\
&1 + 3 + 6 + 10 + \cdots + \frac{n(n+1)}{2} = \frac{n(n+1)(n+2)}{6},\\
&1 + 4 + 10 + 20 + \cdots + \frac{n(n+1)(n+2)}{6} = \frac{n(n+1)(n+2)(n+3)}{24},\\
&1^2 + 2^2 + 3^2 \cdots + n^2 = \frac{n(n+1)(2n+1)}{6}.
\end{aligned}$$

3.4 IMPROVEMENTS IN NOTATION; NAPIER'S INVENTION OF LOGARITHMS

The great advances made in mathematics after the 1600s would not have been possible without an improvement in notation and mathematical language. Two of the major figures in these developments were François Viète (1540–1603) and Simon Stevin (1548–1620).

Although quadratic, and sometimes higher degree polynomial equations, had been studied since the time of the ancients, these had always been in the form of specific equations, treated on a case by case basis. Viète took the decisive step of studying equations with unspecified (arbitrary) coefficients. Thus it became possible for the first time to develop general rules and procedures for a whole class of equations at one stroke. The enormous advantage of this type of generality is obvious just in considering the formula

$$x = \frac{-b \pm \sqrt{b^2 - 4ac}}{2a}$$

for the roots of the quadratic equation

$$ax^2 + bx + c = 0,$$

in comparison for the cumbersome solutions of the Greeks and other ancient civilizations.

Viète did retain the habit of indicating the powers to which a quantity was raised, in language, with terms such as *quadratus*, for power 2, *cubus* for power 3, etc. Mathematics was later freed from this by Descartes, who introduced the modern notation a^n to denote powers. Other advancements in notation also happened in this era; in 1514, the Dutch mathematician Vander Hoecke first used the signs $+$ and $-$ in algebraic expressions (prior to this time these symbols had been used strictly in business, to refer to surpluses and deficits).

In a work entitled *The Whetstone of Witte* (1557), the English mathematician Robert Recorde introduced the symbol $=$ for equality (but with longer lines), chosen on the grounds that "Noe 2 thynges can be moare equalle" than two parallel lines. In 1631, Thomas Harriot introduced the dot as a symbol for multiplication, while another English algebraist William Oughtred introduced the cross sign $\times$ to denote the same operation. (Nowadays it is the usual practice of mathematicians to omit a symbol for multiplication entirely when using the operation in an expression.) In modern notation, the arcane looking statement that Diophantus would have written

$$K^{\gamma}4?^{1}\Delta^{\gamma}6\iota\zeta 2M3$$

is expressed as the equation

$$4x^3 - 6x = 2x + 3.$$

The Belgian mathematician Simon Stevin (1548–1620) is credited with the introduction of decimal numbers into the number system to represent fractions. Prior to 1500, decimal and sexagesimal (base 60) numbers were often used together in one number, with the integer part of the number being represented by a base 10 number and the fractional part in sexagesimal notation. Vestiges of this practice remain to this day in our division of hours into minutes and seconds and the likewise division of angles.

The Scottish nobleman, mathematician and inventor John Napier (1550–1617) introduced logarithms into arithmetic as a means to facilitate calculations with very large numbers. Napier was an interesting character. Living in Scotland at a time of great religious upheaval, he was a vocal defender of the protestant faith. In 1594 Napier published a pamphlet, *A Plaine Discovery on the Whole Revelation of Saint John* in which he bitterly attacked the Catholic church, proclaiming the Pope to be the anti-Christ. The "Discovery" in the title is the finding that the Creator proposed to end the world sometime between 1688 and 1700. The pamphlet received wide circulation, as a result of which Napier became in his day much more famous as a polemicist than he ever was as a mathematician.

Returning to the earthly realm, the term *logarithm* coined by Napier, derives from the Greek *logos* meaning "reckoning." In studying tables of powers, Napier realized that calculating the product of two large numbers could be

[1]?: Here goes a symbol that I don't even recognize sufficiently to typeset!

done much more easily via their logarithms. The logarithm of a number x to a base b is defined as the exponent y such that $b^y = x$, and is denoted $\log_b x$. (As examples, $\log_5 25 = 2$ since $5^2 = 25$ and $\log_{10} 0.1 = -1$ since $10^{-1} = 0.1$.) It follows from the properties of exponents that logarithms obey the addition rule

$$\log_b M + \log_b N = \log_b MN.$$

Suppose, for example, it is required to calculate the product of the two numbers $M = 15346$ and $N = 279435$, and a table of logarithms to base 10 (*common* logarithms) is at hand. The values $\log_{10} M = 4.185995$ and $\log_{10} N = 5.446280$ imply

$$\log_{10} MN = 4.1865995 + 5.446280 = 9.633275.$$

The table then yields $MN = 4288207595$ as the answer. (The actual value is 4288209510, the discrepancy being due to rounding error.)

The use of logarithms thus turns a rather tedious multiplication into a much easier addition.

Exercise 3.8. *Use logarithms to calculate* $1234^2 \times 567^3$.

Napier devoted the latter part of his career to calculating an extensive table of logarithms, using $10^7 - 1 = 9999999$ as the base. Logarithms to this base are called *Naperian* logarithms in his honor. Napier compiled his table by means of an ingenious interpolation method that he devised, whereby known logarithms of two numbers on either side of a number are used to estimate the logarithm of the number. (This is Napier's reason for working with such a strange base.) Napier also invented a mechanical device known as *Napier's bones* to facilitate calculations with logarithms, a forerunner of the modern (now obsolete!) slide rule.

3.5 DESCARTES AND THE INVENTION OF ANALYTIC GEOMETRY

René Descartes (1596–1650) was born in the town of La Haye, about 200 miles southwest of Paris. His father was a member of the lesser nobility and a provincial judge. In 1617, when Descartes was 21 years of age, he enlisted in the French army as a gentleman volunteer, serving first under Prince Maurice of Nassau in Holland and then under the Duke of Brunswick, but there is no evidence of his ever having done any real soldiering. Instead, he concentrated on philosophy and mathematics.

The theory that laid the foundations of analytic geometry first appeared as the third of three essays titled *La Géométrie* accompanying the *Discours*, a philosophical work published by Descartes in 1637. *La Géométrie* itself consisted of three books. In this work Descartes conceived the idea of representing points in a plane by two coordinates with respect to fixed axes, and describing

FIGURE 3.7 Portrait of Descartes by Frans Hals

curves in the plane by means of an equation relating these coordinates. The importance of this innovation for the future of mathematics cannot be overstated. Geometry was finally freed from its Euclidean shackles! An infinitude of new curves could be created at will and studied by means of algebra. Indeed, some commentators herald this work of Descartes as the beginning of modern mathematics. Analytic geometry paved the way for the invention of calculus and a host of other developments in mathematics and related branches of science.

The problem that formed the central theme of *La Géométrie* was classical in nature, a generalization of a problem that Pappus had formulated in his commentary on the conics of Apollonius. The problem was as follows: given four lines in the plane, find the locus of a point that moves so that the product of the distances from two of the lines is proportional to the distance from a third line. Pappus had stated that the locus was one of the conic sections and Descartes was able to prove this algebraically using his new methodology.

In Descartes' own words: *I would simplify matters by considering one of the given lines and one of those to be drawn as the principal lines to which I shall try to refer all others.*

Descartes' coordinate system is illustrated in figure 3.8. In this figure C is considered to be an arbitrary point whose locus is to be determined according to the rules of the problem. Descartes chooses one of the lines specified in

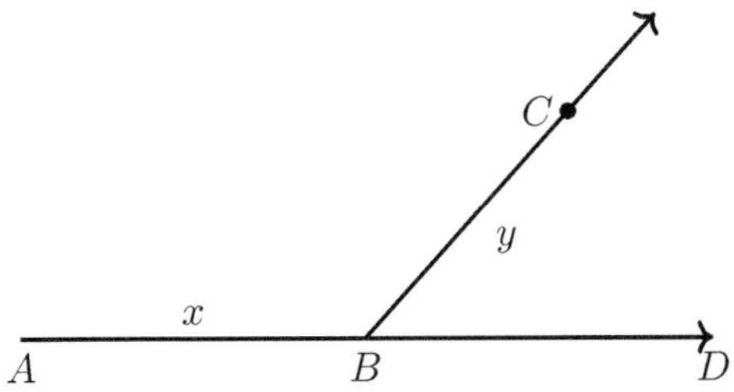

FIGURE 3.8 Descartes' coordinate system

the problem (the line AD in the figure) as one of his axes, then draws a line through C in a certain fixed direction and projects the line until it meets AD at the point B. The location of the point C is represented by the lengths x and y of the line segments $|AB|$ and $|BC|$. Descartes sets up an equation in x and y and uses it to determine locus of the point C.[2]

In Book II of *La Géométrie*, titled *On the Nature of Curved Lines*, Descartes enlarged on this approach and divided the curves produced by his new geometry into two types which he called *geometric* and *mechanical* (later termed *algebraic* and *transcendental* by Leibniz). By the former, he meant the curves defined by an algebraic equation in two variables. Curves such as spirals, which are not of this type, were classified by Descartes as mechanical.

Descartes also gave a method for constructing tangent lines to curves described by an algebraic equation $f(x, y) = 0$. Descartes was motivated to study this problem by his researches in optics. He himself said of the tangent problem: *I dare say that this is not only the most useful and the most general problem in geometry that I know, but even that I have ever desired to know.* No essential progress had been made on this problem since Archimedes, who calculated tangents to his spirals.

As an example of Descartes' method of tangents, consider the parabola $y^2 = 2x$. Suppose we wish to construct the tangent line to this curve at the point $P = (8, 4)$. The problem is clearly equivalent to constructing the normal line to the curve at P. Descartes does this by calculating the circle centered on the x-axis, passing through P, that is tangential to the parabola at P (see figure 3.9).

The procedure is as follows. Denote the center of the circle as the point $(x_1, 0)$, so the equation of the circle through P is $(x-x_1)^2+y^2 = (8-x_1)^2+16$. In order for the circle and the parabola to be tangent at P, the simultaneous

[2]Descartes' presentation of these ideas was a little complicated by the facts that his axes were not assumed to be perpendicular, as we do today.

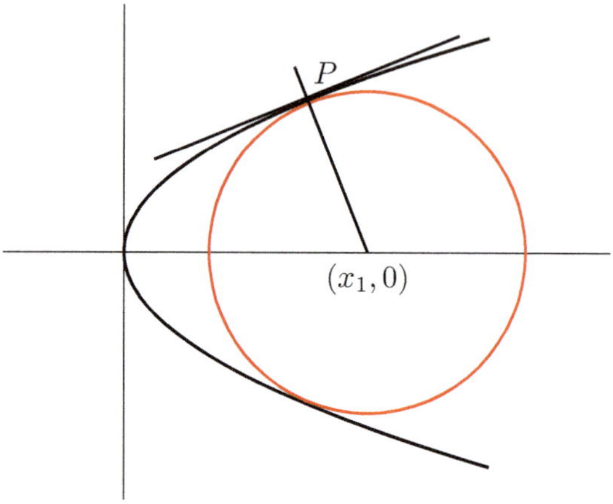

FIGURE 3.9 Descartes' method of tangents

equations

$$(x - x_1)^2 + y^2 = (8 - x_1)^2 + 16,$$
$$y^2 = 2x$$

determining the intersection of the two curves, must produce a *single solution point*. Eliminating y between the equations yields

$$(x - x_1)^2 + 2x = (8 - x_1)^2 + 16.$$

Write this equation in the form

$$x^2 + Bx + C = 0,$$

where $B = -2x_1 + 2,\ \ C = 16x_1 - 80$. The quadratic equation admits a single solution if and only if $B^2 - 4C = 0$. This is easily shown to yield $x_1 = 9$. Hence, the normal line to the parabola at the point $(8, 2)$ is the radius of the circle centered at (9,0) connecting to the point $(8, 2)$. The tangent line can then be constructed as the line perpendicular to this line at the point in question.

If, in retrospect, this all seems relatively straightforward, it should be noted that such an argument could not even be conceived of without the analytic geometry previously introduced by Descartes.

Book III of *La Géométrie* concerns purely algebraic problems. Here Descartes gives his famous "rule of signs" which provides an upper bound on the number of positive real roots to a polynomial equation in terms of the number of sign changes in the coefficients of the equation. In this book, Descartes also gives a new solution to the quartic equation.

As usual, Descartes treats the equation in its reduced form

$$z^4 + pz^2 + qz + r = 0.$$

The left-hand side is expressed as the product of two quadratic terms

$$(z^2 + kz + m)(z^2 - kz + n) =$$
$$z^4 + (m + n - k^2)z^2 + k(n - m)z + mn.$$

Comparing the coefficients in the two forms of the equations leads to the relations

$$p = m + n - k^2, \quad q = k(n - m), \quad r = mn.$$

If $k = 0$, then $q = 0$ and the original equation is a quadratic in z^2, which can be solved. If $k \neq 0$,then the first two relations above give

$$2n = p + k^2 + \frac{q}{k}, \quad 2m = p + k^2 - \frac{q}{k}.$$

Substituting these in the third relation yields

$$k^6 + 2pk^4 + (p^2 - 4r)k^2 - q^2 = 0.$$

This is a cubic equation in k^2, which is then solved by Cardano's formula. Once k, and hence n and m have been found, the roots z are obtained by solving the pair of quadratic equations

$$z^2 + kz + m = 0,$$
$$z^2 - kz + n = 0.$$

3.6 PRECURSORS TO CALCULUS: WALLIS AND BARROW

The Englishman John Wallis (1616–1703) was probably the most talented mathematician to emerge in the period between Descartes and Newton. In 1648 Wallis was appointed Savilian Professor of Geometry at Oxford University, a position he held until his death. Wallis' mathematical reputation is largely based on a treatise, *Arithmetica Infinitorum* which he published in 1655. In this book, following up on earlier work of the Italian mathematician Cavalieri, (1598–1647), Wallis developed a method for computing the area beneath the curves $y = x^k$ for k positive integers k. This work inspired both Newton and Leibniz in their later creation of the integral calculus.

As an illustration of Wallis' method, consider the parabolic region $y = x^2$, $x \leq a$. Wallis supposed the region to be comprised of n infinitesimally narrow rectangles, of width a/n (see figure 3.10).

Wallis reasoned that the total area A of the region is the sum of the areas of the rectangles,

$$\left(\frac{a}{n}\right)^2 \cdot \frac{a}{n} + \left(\frac{2a}{n}\right)^2 \cdot \frac{a}{n} + \left(\frac{3a}{n}\right)^2 \cdot \frac{a}{n} + \cdots + \left(\frac{na}{n}\right)^2 \cdot \frac{a}{n}$$
$$= \frac{a^3}{n^3}\left(1^2 + 2^2 + 3^2 + \cdots + n^2\right).$$

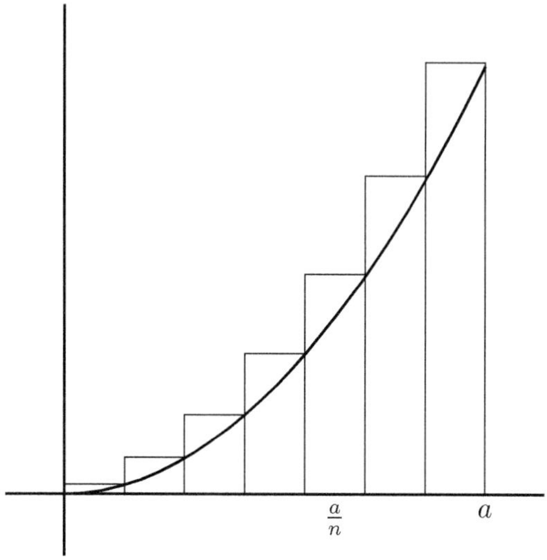

FIGURE 3.10 Computing the area beneath a curve

Wallis had observed the pattern

$$\frac{1^2}{1^2+1^2} = \frac{1}{2} = \frac{1}{3} + \frac{1}{6},$$
$$\frac{1^2+2^2}{2^2+2^2+2^2} = \frac{5}{12} = \frac{1}{3} + \frac{1}{12},$$
$$\frac{1^2+2^2+3^2}{3^2+3^2+3^2+3^2} = \frac{7}{18} = \frac{1}{3} + \frac{1}{18},$$
$$\frac{1^2+2^2+3^2+4^2}{4^2+4^2+4^2+4^2+4^2} = \frac{3}{8} = \frac{1}{3} + \frac{1}{24}$$

from which he deduced

$$A = a^3\left(\frac{1}{3} + \frac{1}{2n} + \frac{1}{6n^2}\right).$$

Since n is assumed to be infinite, the terms $1/2n$ and $1/6n^2$ are zero, hence $A = a^3/3$.

Wallis performed an analogous calculation with the curve $y = x^3$ and found that the corresponding area is $a^4/4$.

Exercise 3.9. *Use the summation formula*

$$1^3 + 2^3 + 3^3 + \cdots + n^3 = \frac{n^2(n+1)^2}{4}$$

to establish this.

On the basis of these and further results, Wallis came to the conclusion that for every positive power k, the area lying beneath the curve $y = x^k$ is

$$\frac{a^{k+1}}{k+1}.$$

As we have remarked, and will discuss in greater detail in Chapter 5, the work of Cavalieri and then Wallis on this problem inspired the invention of integral calculus by Newton and Leibniz.

Barrow's work on computing tangents to curves played a similar role in the formulation of that other major branch of the subject, differential calculus. Isaac Barrow (1630–1677) arrived in Cambridge in 1663 as the first occupant of the Lucasian Chair of Mathematics, when Newton was a student there. Like Wallis, Barrow was one of the foremost mathematicians of his time. His influence on Newton was considerable and apparently inspired the younger man to follow an academic career. Barrow's researches into curves led him to the very brink of discovering the differential side of calculus.

During his tenure at Cambridge, Barrow presented a series of 13 lectures. These were published in 1669 under the name *Lectiones Geometricae* and contained, among other things, Barrow's method of tangents. (Fermat independently came up with similar ideas.) This method was later taken up by Newton and is essentially the one we use today for calculating the derivative. The method can be described as follows. Suppose the curve is defined by an equation $f(x, y) = 0$ where f is a polynomial in x and y. Barrow observed that the tangent line at a point $P = (x, y)$ on the curve is determined if some other point on the line is found, say the point T where the tangent line meets the horizontal axis (see figure 3.11).

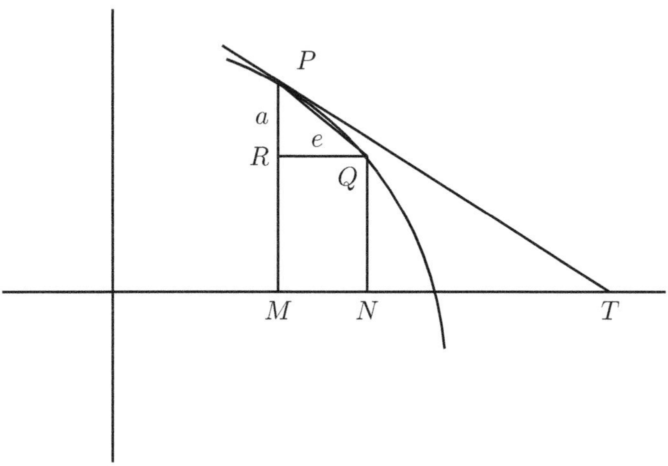

FIGURE 3.11 Barrow's method of tangents

Barrow considers a point Q on the curve infinitesimally close to P, and constructs the right triangle PQR which he calls the "differential triangle." The closer the point Q is to P, the more similar are the triangles PRQ and PTM. Barrow argued that they can actually be assumed to be similar. This implies the relation

$$\frac{MP}{MT} = \frac{RP}{RQ} = -\frac{a}{e}. \tag{3.6}$$

Taking (x, y) as the coordinates of P, the coordinates of Q (lying below and to the right of P in figure 3.11) are $(x + e, y - a)$. Since Q lies on the curve, we have

$$f(x + e, y - a) = 0.$$

The left-hand side of the equation is expanded as a polynomial in a and e. In Barrow's own words: *reject all terms in which there is no a or e, for they destroy each other by the nature of the curve*[3]*; also reject all terms in which a and e are above the first power, or are multiplied together, for they are of no value with the rest, being infinitely small(er).* This results in a linear relation in a and e from which one can solve for a/e. By (3.6), this yields the slope of the tangent line at P and hence the point T.

Let's try this for the curve

$$f(x, y) = x^2y + y^3 + 2x^2 - 5 = 0.$$

Let $P = (x, y)$ be an arbitrary pint on the curve. Taking a nearby point $Q = (x + e, x - a)$ on the curve, we have

$$0 = f(x + e, y - a) = (x + e)^2(y - a) + (y - a)^3 + 2(x + e)^2 - 5.$$

Expanding the square terms and effecting the further multiplications, yields

$$\begin{aligned} 0 &= (x^2 + 2xe + e^2)(y - a) + y^3 - 3y^2a + 3ya^2 - a^3 + 2(x^2 + 2xe + e^2) - 5 \\ &= x^2y - x^2a + 2xey - 2xea + e^2y - e^2a + y^3 - 3y^2a + 3ya^2 - a^3 \\ &+ 2x^2 + 4xe - 2e^2 - 5. \end{aligned}$$

Canceling out the combination of terms $x^2y + y^3 + 2x^2 - 5$ and ignoring terms in a and e of higher degree than first, results in

$$a(x^2 + 3y^2) - e(2xy + 4x) = 0,$$

from which we get

$$\frac{a}{e} = \frac{2xy + 4x}{x^2 + 3y^2}.$$

Hence the slope of the tangent line at (x, y) is

$$-\frac{a}{e} = -\frac{2xy + 4x}{x^2 + 3y^2}$$

[3] According to the relation $f(x, y) = 0$.

and the point T has x-coordinate

$$x + y \cdot \frac{2xy + 4x}{x^2 + 3y^2}.$$

Those who have taken a first course in calculus may recognize what we have done in this calculation as performing an implicit differentiation with respect to x in the relation $f(x, y) = 0$.

Exercise 3.10. *Use Barrow's method to find the equation of the tangent line to the curve described by the following relation, at the point* $(1, 2)$

$$3x^2y + 2y^2x = 14.$$

This was the state of (what came to be called) calculus in the mid to late 1600s. Methods had been introduced to solve two fundamental problems in geometry, the calculation of areas and tangent lines. However, the two types of problems were treated by Wallis and Barrow and their contemporaries in isolation, and on a case-by-case basis. Some formulas, but no general theory had been established from the calculations. The subject was soon to undergo a transformation in the hands of two men of genius that would place it forever at the forefront of mathematical thinking.

Suggestion for Further Reading

David M. Burton. *The History of Mathematics: An Introduction.* McGraw Hill, 7th edition, 2010.

CHAPTER 4

Algebra

The focus of this chapter is on polynomial equations, a driving theme in algebra throughout the ages and the source of much deep and beautiful mathematics.

The chapter starts with a review of complex numbers and moves on to discuss the Fundamental Theorem of Algebra (FTA). This is the assertion that a polynomial equation with real or complex coefficients has as many complex roots as the degree of the equation. The next three sections of the chapter discuss, in turn, quadratic, cubic, and quartic equations. We present the standard methods to solve these equations and relate something of the history of these methods. (The reader will find some of the history quite colorful.) Section 4.6 gives an account of de Moivre's theorem, an elementary but important result which expresses the nth roots of unity, i.e., the solutions to the equation $z^n = 1$, in the form of sines and cosines.

After the solution to the quartic equation was discovered around the middle of the sixteenth century, mathematicians naturally set their sights on finding algebraic formulas to solve polynomial equations of degree 5 and higher. However, no definitive progress in this direction was made for another 300 years, when Ruffini and then Abel (more convincingly) resolved the issue in the negative, proving that no such formulas can exist. The theoretical basis for this research was laid by Lagrange, who in 1770 published an influential memoir on the subject of polynomial equations. We discuss Lagrange's memoir in Section 4.7, together with a work of Vandermonde on the same theme which, in addition to many of the results of Lagrange, contains a revolutionary new approach to the cyclotomic equation of degree 11. In Section 4.8, we describe Gauss's Euclidean construction of the regular 17-sided polygon.

The crowning glory of the above lines of research is the theory pioneered by Galois in the early 1800s, which provides the complete answer to the question of the algebraic solvability of polynomial equations. The chapter concludes with an introduction to Galois theory and its relation to the work of Ruffini and Abel.

DOI: 10.1201/9781003470915-4

4.1 COMPLEX NUMBERS

It may seem strange to begin a chapter on Algebra with a discussion of complex numbers, but they are clearly fundamental to the subject. Indeed, the need for complex numbers arises from quadratic equations. Take, for example,

$$x^2 + 4 = 0.$$

Since the rules of arithmetic dictate that when either a positive or a negative number is squared, the result is always non-negative, it follows that no (real) number can satisfy this equation. The quadratic formula, which gives for the equation $x^2 + px + q = 0$ the solutions

$$y = \frac{-p \pm \sqrt{p^2 - 4q}}{2}$$

shows that this problem will occur whenever the discriminant $p^2 - 4q$ of the equation is negative.

Many of the ancient civilizations studied quadratic equations and possessed a version of the quadratic formula, though expressed verbally and usually in relation to specific problems. They must have come across equations such as these. This was not seen as a problem in those days; such equations were simply declared to have no solutions, or "impossible."

The real impetus for the introduction of complex numbers into mathematics came in the sixteenth century when algebraic solutions to cubic equations were discovered by Tartaglia and del Ferro. It turns out that complex numbers cannot be avoided in these solutions, even in cases when all three roots are real.

The term "imaginary" for the square roots of negative numbers (which does not exactly inspire confidence) was coined by Descartes in 1637 in the statement, ... *sometimes only imaginary, that is one can imagine as many as I said in each equation, but sometimes there exists no quantity that matches that which we imagine.* Even the great Gauss, who introduced the symbol "i" for $\sqrt{-1}$ and the terms "argument" and "modulus" of complex numbers, initially distrusted them and referred in 1797 to the "metaphysics of the square root of -1."

Complex numbers had, in fact, been used much before this time and had been seen to produce meaningful results. Arithmetic rules for their manipulation were formulated by Raphael Bombelli as early as 1572. Yet the question persisted, *do such numbers actually exist?* The matter was put to rest by the Danish mathematician Caspar Wessel who, in a memoir of 1799 devised the idea of representing complex numbers as points in a plane. Wessel's memoir was published in the *Proceedings of the Copenhagen Academy*, but went largely unnoticed in the mathematical world. Similar ideas were formulated independently by Jean-Robert Argand who, in 1806, used complex numbers to give what is considered to be the first rigorous proof of the fundamental theorem of algebra. The theory of functions of a complex variable was developed by Augustin-Louis Cauchy (1789–1857), and Bernhard Riemann (1826–1866) in the mid-nineteenth century. All these developments will be discussed in this chapter.

FIGURE 4.1 Augustin-Louis Cauchy

FIGURE 4.2 Bernhard Riemann

Complex numbers are today an essential part of mathematics and are used in many scientific disciplines. Examples include electrical engineering, fluid dynamics, signal analysis, and quantum physics. If you are looking to impress people at parties with complex numbers, the best way to describe them is as "the quotient ring of the polynomial ring in the indeterminate i, by the ideal generated by the polynomial $i^2 + 1$."

The complex plane

Many people come across complex numbers in a high school math class. The teacher will usually introduce the subject by saying something like: "Since there is no number that satisfies $x^2 = -1$, mathematicians have invented one and they call this new number i. This leads to complex numbers such as $3+4i$ and $2-5i$. You can add these numbers, subtract them, multiply and divide them, just as you do with regular numbers." For example, "foiling" the product gives

$$\begin{aligned}(3+4i)(2-5i) &= (3\times 2) + (3\times -5i) + (4i\times 2) + (4i\times -5i)\\ &= 6 - 15i + 8i - 20i^2 = 6 - 7i - 20i^2.\end{aligned}$$

Since $i^2 = -1$, this can be written as

$$6 - 7i + (-20\times -1) = 6 - 7i + 20 = 26 - 7i.$$

All well and good, but is there any mathematical basis for these new numbers? The issue was especially baffling to the mathematical world in the late 1700s. What was needed was some sort of tangible construction to justify the existence of imaginary and complex numbers. This is exactly what Wessel and Argand provided by representing complex numbers as points in a plane.

The complex plane is shown in figure 4.3. The horizontal axis (labeled Re for real) in the figure is a (real) number line. The vertical axis (labeled Im for imaginary) is also a number line, but is calibrated in terms of a unit i. Points (x, y) in the plane are written $x + yi$ (or $x + iy$), called *complex numbers*, and often denoted by a single letter z.

Addition and multiplication of complex numbers are formally defined by the rules

$$\begin{aligned}(a+bi)\oplus(c+di) &= (a+c)\pm(b+d)i,\\ (a+bi)\otimes(c+di) &= (ac-bd)+(ad+bc)i.\end{aligned}$$

Note, that in the case when the two numbers are real (i.e., $b = d = 0$), we have $a\oplus b = a+b$ and $a\otimes b = ab$. Also, $i\otimes i = -1$. The real numbers are hereby embedded in a larger arithmetical framework which is given a visual reality, and an element i is produced whose square is -1.

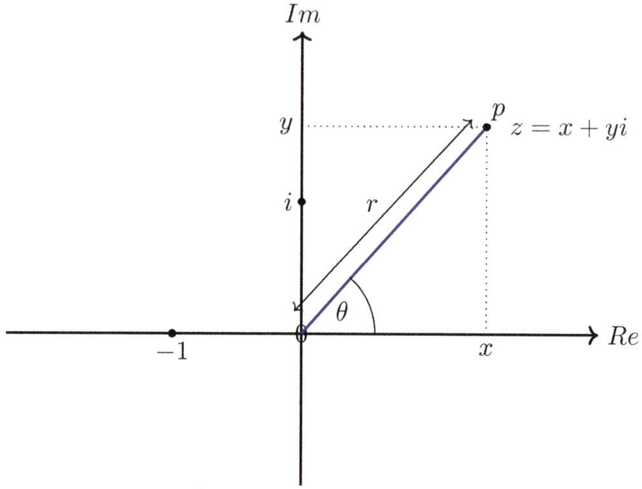

FIGURE 4.3 The complex plane

Modulus and Argument

The length r of the line segment $0p$ is called the *modulus* of z and denoted $|z|$. The angle θ between $0p$ and the positive real axis (measured in the counter-clockwise sense) is called the *argument* of z and denoted $\arg z$. Thus

$$|z| = \sqrt{x^2 + y^2},$$

$$\arg(z) = \arctan\left(\frac{y}{x}\right).$$

Since $x = r\cos(\theta)$ and $y = r\sin(\theta)$, we can write

$$z = |z|\left(\cos(\arg(z)) + i\left(\sin(\arg(z))\right)\right).$$

Addition $\oplus$, as defined above, has the geometric interpretation as the "parallelogram law" shown in figure 4.4.

The geometric meaning of the multiplication law $\otimes$ is indicated by the properties

$$|z \otimes w| = (|z|)(|w|), \tag{4.1}$$

$$\arg(z \otimes w) = \arg(z) + \arg(w). \tag{4.2}$$

Proof. Let $z = (a + bi), w = (c + di)$. Then

$$(|z|)(|w|) = \sqrt{a^2 + b^2}\,\sqrt{c^2 + d^2} = \sqrt{(ac - bd)^2 + (ad + bc)^2} = |z \otimes w|$$

proving (4.1).

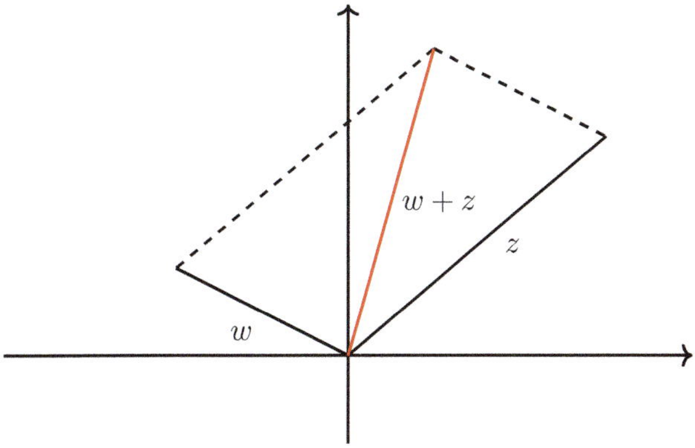

FIGURE 4.4 The parallelogram law for addition

By the addition formula for tangent,

$$\tan(\arg(z) + \arg(w)) = \frac{\tan(\arg(z)) + \tan(\arg(w))}{1 - \tan(\arg(z))\tan(\arg(w))}$$
$$= \frac{b/a + d/c}{1 - (b/a)(d/c)} = \frac{ad + bc}{ac - bd} = \tan\left(\arg(z \otimes w)\right).$$

This proves (4.2). □

In particular, (4.1) and (4.2) imply $i^2 = i \times i$ has modulus $1^2 = 1$ and argument $\pi/2 + \pi/2 = \pi$, i.e., $i^2 = -1$.

(The symbols $\oplus$ and $\otimes$ have been used to emphasize the fact that, at this point, these were introduced as formal operations, divorced from their usual arithmetic meanings. These symbols will now be dropped and the operations denoted by the familiar + and ×.)

4.2 THE FUNDAMENTAL THEOREM OF ALGEBRA

The Fundamental Theorem of Algebra (FTA) asserts that polynomial equations with (real or) complex coefficients, can be solved by complex numbers. A precise statement of the theorem is as follows.

Theorem 4.1. *(FTA) The polynomial equation*

$$p(z) = a_n z^n + a_{n-1} z^{n-1} + \cdots + a_0 = 0, \tag{4.3}$$

where $n \geq 1$, and $a_0, a_1, \ldots, a_n$ are complex numbers with $a_n \neq 0$, has at least one complex solution (root).

An equivalent (seemingly stronger) statement of the theorem is the following.

Theorem 4.2. *Equation 4.3 has exactly n complex roots, up to multiplicity (some may be repeated).*

To see that Theorem 4.2 follows from Theorem 4.1, suppose that z_1 is a solution of equation (4.3) as provided for by the theorem. If $n = 1$, then there is nothing more to prove. If $n > 1$, then dividing p through by $z - z_1$ produces the factorization

$$p(z) = (z - z_1)q(z),$$

where q is a polynomial of degree $n - 1$. Then, again by Theorem 4.1, q has a root, z_2. Continue this process until n roots have been obtained.

Here is yet another equivalent version of FTA.

Theorem 4.3. *Every polynomial equation with real coefficients can be expressed as a product of linear and quadratic factors with real coefficients.*

Exercise 4.1⋆ *Derive Theorem 4.3 from Theorem 4.2.*

FTA has a long history. In a book of 1629, Albert Girard stated that a polynomial equation of degree n has n solutions, but did not say anything about the nature of the solutions. The theorem (or one of its equivalent versions) was not always believed. Indeed, no lesser an authority than Leibniz stated in 1702 that the polynomial $x^4 + a^4$ could not be written as the product of two quadratic factors. This was later disproved by Newton, who obtained the factorization

$$x^4 + a^4 = (x^2 - \sqrt{2}ax + a^2)(x^2 + \sqrt{2}ax + a^2).$$

A similar claim was made about a more complicated expression by Nikolaus Bernoulli, and this was subsequently disproved by Euler.

Over time, it was realized that the theorem had to be true, in some form, and several mathematicians attempted to prove it, including Euler (1749), de Foncenex (1759), Lagrange (1772), and Laplace (1795). All these proofs contained gaps or flaws. Gauss gave six different proofs of the theorem, although all are regarded as incomplete by today's standards. The first rigorous proof of FTA is credited to Argand, in 1806.

We sketch one of the many proofs of FTA. Denote the polynomial by

$$p(z) = a_n z^n + a_{n-1} z^{n-1} + \cdots + a_0, \tag{4.4}$$

where $n \geq 1$ and $a_n \neq 0$. We can assume $a_0 \neq 0$, otherwise $z = 0$ is a root of (4.4) and there is nothing further to prove.

Consider the family of circles in the complex plane

$$\sigma_r = \{r(\cos t + i \sin t),\ 0 \leq t \leq 2\pi\},$$

(which, for conciseness, we write as σ_r), and think of these as paths beginning and ending at the point $z = r$. We study the images of the paths under the map p

$$p(\sigma_r) = \{p(z), z \in \sigma_r\}.$$

Two such paths and their images are depicted in figure 4.5. The key points in the argument are the following:

1. Since the paths σ_r are closed loops, i.e., the beginning and end points coincide, the same is true of $p(\sigma_r)$.

2. For $|z|$ small, $p(z)$ behaves like $p_1(z) = a_0 + a_1 z$ (since the contribution of the higher-order terms is negligible). The path $p_1(\sigma_r)$ is a circle centered at a_0.

3. For large $|z|$, the highest-order term in the polynomial dominates, and $p(z)$ behaves like $p_n(z) = a_n z^n$. The path $p_n(\sigma_r)$ is a circle centered at 0, which wraps n-times around 0.

4. In view of points 2 and 3, for small r, the loop $p(\sigma_r)$ (which stays close to $p_1(\sigma_r)$), together with its interior, avoid 0, while for large r, $p(\sigma_r)$ (which stays close to $p_n(\sigma_r)$), wraps n-times around 0. (See the red and blue loops on the right side of figure 4.5.)

Now suppose we let r steadily increase from small to large. Then the loops on the right of figure 4.5 undergo a continuous progression from the red loop, which does not wind around 0) to the blue loop, which winds around 0. This implies that, at some point, the loops must pass over 0, i.e., there exist z such that $p(z) = 0$.

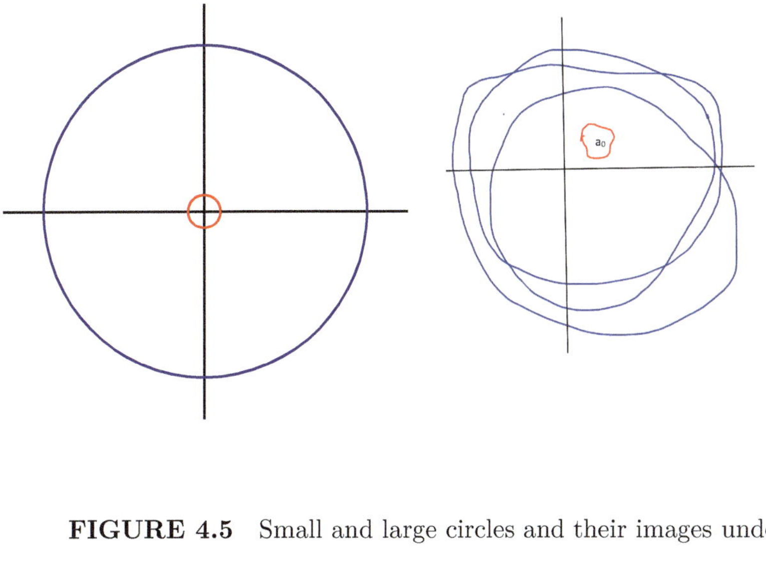

FIGURE 4.5 Small and large circles and their images under p

4.3 QUADRATIC EQUATIONS

If the Pythagorean theorem is the most well-known result in mathematics, the quadratic formula probably comes in a close second. Most high school students learn the quadratic formula as the statement: the equation

$$ax^2 + bx + c = 0$$

has as its solutions

$$x = \frac{-b \pm \sqrt{b^2 - 4ac}}{2a}. \tag{4.5}$$

Quadratic equations appear in the works of most of the ancient civilizations, Greek, Babylonian, Arabic, Indian, Chinese ..., although, not of course expressed in the above terms (there existing then no algebraic language to express either the equation or the solution). For example, a Babylonian tablet dated circa 2000 BCE contains the problem:

I have added the area and two-thirds of the side of my square and it is 0;35. What is the side of my square?

The solution is given in verbal form as follows:

You take 1. Two-thirds of 1 is 0;40. Half of this, 0;20, you multiply by 0;20 and you add to 0;35 and the result 0;41,40 has 0;50 as its square root. The 0;20, which you have multiplied by itself, you subtract from 0;50, and 0;30 is the side of the square.

If this seems insanely cryptic, it is because the Babylonians used a *sexagesimal* (base 60) system of numbers. Thus 0;20 represents the number 20/60 = 1/3, 0;35 represents 35/60 = 7/12, etc. In modern terms, the text gives the solution to the equation

$$x^2 + \frac{2x}{3} = \frac{7}{12}:$$

$$x = \sqrt{\left(\frac{2/3}{2}\right)^2 + \frac{7}{12}} - \frac{2/3}{3} = \frac{1}{2}.$$

Clearly, a recipe is given here for solving the general quadratic equation

$$x^2 + px = q,$$

$$x = \sqrt{\frac{p^2}{4} + q} - \frac{p}{2}.$$

The Babylonians did not describe the methods by which they arrived at their solution to quadratic equations, but, later, al-Khwarizmi did. His method was the modern technique of "completing the square," interpreted literally. al-Khwarizmi solves the quadratic equation

$$x^2 + 10x = 39$$

via the geometric construction shown in figure 4.6. al-Khwarizmi adds the gray square to the figure. This leads to the equation

$$(x+5)^2 = x^2 + 10x + 25 = 39 + 25 = 64$$

from which one readily finds that $x = 3$.

Exercise 4.2. *Use the method of completing the square to derive the quadratic formula (4.5).*

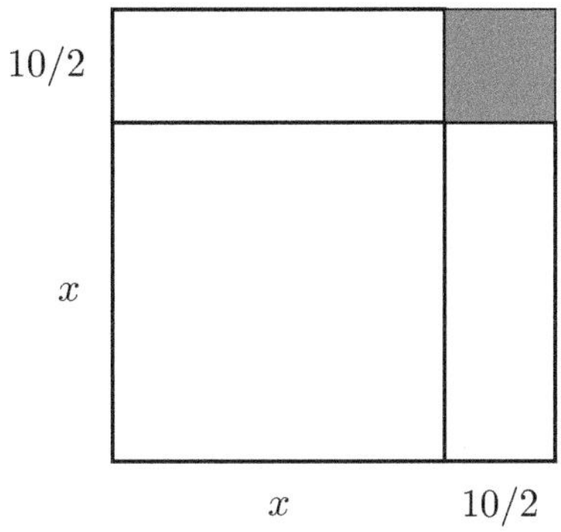

FIGURE 4.6 Completing the square

Before we leave quadratic equations, we should mention a further method for their solution, which is also relevant in solving higher degree equations. This goes by the name of a *Tschirnhaus transformation*, named for a German mathematician Ehrenfried Walther von Tschirnhaus (1651–1708), who introduced it as part of a larger work on polynomial equations. The idea is to make a substitution $x = y + p$ in the equation

$$ax^2 + bx + c = 0$$

then choose p such that the second-highest power of x in the equation vanishes. The substitution leads to the equation

$$a(y+p)^2 + b(y+p) + c = ay^2 + (2ap+b)y + ap^2 + bp + c = 0.$$

Choosing $p = -b/2a$ in the equation reduces it to

$$ay^2 + c - \frac{b^2}{4a} = 0$$

from which (4.5) easily follows.

4.4 CUBIC EQUATIONS

As discussed previously, the solution to quadratic equations was known in antiquity. It is remarkable, then, that it took humankind a further three and a half thousand years to come up with an algebraic solution to the cubic equation

$$ax^3 + bx^2 + cx + d = 0. \tag{4.6}$$

Even more remarkable is that fact when the solution to the cubic was finally discovered, it was found independently and at approximately the same time by two mathematicians, both Italian, Scipione del Ferro (1465–1526) and Niccolò Tartaglia (1500–1557). It was customary at that time for mathematicians to engage in public contests for monetary prizes. The winner would be rewarded by fame and fortune, and the loser with the likely loss of their position and livelihood. Thus if somebody was in possession of such a mathematical gem as the solution to cubic equations, it was prudent to keep it under their hat, publish *and* perish! Tartaglia was persuaded to reveal his solution of the cubic to Cardano, a leading figure in the mathematical world, on the understanding that Cardano keep it a secret. Cardano then published the solution in his famous book on algebra, *Ars Magna*, (The Great Art, 1545) resulting in much acrimony between the two men.

Girolamo Cardano was born in 1501 in the town of Pavia, Italy, the illegitimate child of Fazio, a jurist and a close friend of Leonardo Da Vinci. Cardano had a difficult youth owing to the early departure of his mother and a rough upbringing by an overbearing father, and he grew up to become a combative, though intellectually gifted young man. Cardano graduated with a degree in medicine from the university of Padua in 1525 and after being denied admittance to the College of Physicians in Milan (the sanctioning body for doctors in Italy at that time), on account of his illegitimate birth, supported himself by gambling and the unlicensed practice of medicine. In 1531, he married Lucia Banderini, and they had three children, Giovanni Battista (1534), Chiara (1537), and Aldo Urbano (1543). With the help of some noblemen friends, Cardano finally obtained his medical license and a teaching position in Milan, where he practiced as a doctor and at the same time pursued research into mathematics. He went on to build considerable reputations in both fields.

In retrospect, Cardano's behavior in the matter of the cubic seems fairly honorable. Firstly, *Ars Magna* was published six years after Cardano learned of Tartaglia's solution, thus affording Tartaglia plenty of time to publish the work himself. In the meantime, Cardano had learned of del Ferro's solution and felt that it released him from his pledge to Tartaglia, on the grounds that it was primarily del Ferro's work that he was publishing. Furthermore, a solution to the *quartic* equation had since been discovered in 1540 by Cardano's student, Ludovico Ferrari. Cardano wanted to include Ferrari's solution in the book, but was unable to so without also revealing the work of Tartaglia or del Ferro, since the solution of the quartic relies upon that of the cubic. Perhaps most importantly, Cardano credited both Tartaglia and del Ferro as the discoverers of the solution and so helped secure their place in mathematical history. (True, the formula for the solution did come to be named after Cardano rather than its originators, but that's a story for another day.)

FIGURE 4.7 Nicolo Tartaglia

The algebraic solution to the cubic equation goes as follows. We start by assuming the coefficient $a = 1$ in equation 4.6 (otherwise divide through by it). The Tschirnhaus transformation $x = y - b/3$ reduces the equation to the "depressed" form

$$y^3 + py + q = 0, \tag{4.7}$$

where

$$p = \frac{-b^2}{3} + c, \quad q = \frac{-b^3}{27} + \frac{b^2}{9} - \frac{bc}{3} + d.$$

Substituting $y = u + v$ into (4.7), we have

$$(u+v)^3 + p(u+v) + q = 0.$$

Expanding the cubic binomial and grouping terms yields

$$u^3 + v^3 + q + 3uv(u+v) = 0. \tag{4.8}$$

We separate this into the two equations

$$y^3 + py + q = 0. \tag{4.9}$$

FIGURE 4.8 Girolamo Cardano. Stipple engraving by R. Cooper

$$3uv + p = 0. \tag{4.10}$$

Solving equation 4.10 for v and substituting the result into (4.9) yields

$$u^3 - \frac{p^3}{27u^3} + q = 0 \tag{4.11}$$

or

$$27u^6 + 27qu^3 - p^3 = 0. \tag{4.12}$$

Equation 4.12 is a quadratic equation in u^3. Applying the quadratic formula and reducing, we obtain

$$u^3 = \frac{-9q \pm \sqrt{81q^2 + 12p^3}}{18}. \tag{4.13}$$

Thus u can be found, then v, then y, and finally x.

Exercise 4.3. *Use the Tartaglia-del Ferro method to solve the cubic equation*

$$x^3 + 13x - 9 = 0.$$

The Tartaglio-del Ferro method results in the following general solution to the cubic equation 4.7, known as *Cardano's formula*

$$x = \sqrt[3]{-\frac{q}{2} + \sqrt{\frac{q^2}{4} + \frac{p^3}{27}}} + \sqrt[3]{-\frac{q}{2} - \sqrt{\frac{q^2}{4} + \frac{p^3}{27}}}. \tag{4.14}$$

Exercise 4.4.* *Use a Tschirnhaus substitution and then Cardano's formula to find a solution to the cubic equation*

$$x^3 + 3x^2 + 7x + 7 = 0.$$

Exercise 4.5. *Derive a proof of Cardano's formula by performing the following steps:*

(i) Set $x = \sqrt[3]{a+b} + \sqrt[3]{a-b}$ *and calculate* x^3.

(ii) Express your answer in terms of x *and thereby obtain a cubic equation for* x.

(iii) Equate the coefficients of the linear term and the constant term to p *and* q*, respectively.*

(iv) Solve for a *and* b *in terms of* p *and* q.

Cardano's formula raises some puzzling issues. For example, consider the equation

$$x^3 + 3x - 36 = 0. \tag{4.15}$$

Since the function $y = x^3 + 3x - 36$ is strictly increasing, this equation has only one real solution and inspection shows that it is $x = 3$. However, Cardano's formula yields

$$x = \sqrt[3]{18 + \sqrt{325}} + \sqrt[3]{18 - \sqrt{325}}$$

and it is by no means obvious that this equals 3.

An even stranger situation occurs when one applies Cardano's formula to the equation

$$x^3 - 15x - 4 = 0.$$

The formula yields

$$x = \sqrt[3]{2 + \sqrt{-121}} + \sqrt[3]{2 - \sqrt{-121}}.$$

Why does Cardano's formula return an expression involving complex numbers when the equation has the obvious real solution $x = 4$? This example is discussed in *Arts Magna*. It must have seemed very baffling in Cardano's day when not even negative numbers, let alone complex ones, were trusted. Cardano was a savvy enough mathematician to realize that if one assumed that such quantities had meaning and manipulated them as if they were numbers, then it did lead to correct conclusions. Cardano summed up this approach with the somewhat dismal sounding maxim, *progresses algebraic subtlety, the end of which is as refined as it is useless.*

Cardano's formula proved to be the main driving force behind the introduction of complex numbers into mathematics. This seems odd when one considers that the quadratic formula, with its term $\sqrt{b^2 - 4ac}$ seems to point directly to them. In fact, mathematicians of that era were quite willing to declare that quadratic equations with negative discriminant $b^2 - 4ac$, have no solutions. However, when it was realized that, in certain cases (such as the one above), Cardano's formula requires complex numbers to obtain *real* roots of cubic equations, mathematicians felt the need to invent them.

4.5 QUARTIC EQUATIONS

Ferrari's solution to the quartic equation

$$ax^4 + bx^3 + cx + d + e = 0$$

works by solving a derived cubic. Consider the depressed form of the equation

$$y^4 + py^2 + qy + r = 0. \tag{4.16}$$

(Once again, this can be effected by a Tschirnhaus transformation). First, write equation 4.16 as

$$y^4 + py^2 = -qy - r,$$

then complete the square on the left-hand side by adding $py^2 + p^2$ to each side, This yields

$$(y^2 + p)^2 = py^2 + p^2 - qy - r.$$

Now add a term $2(y^2 + p)z + z^2$ onto each side to make the left-hand side of the form $(y^2 + p + z)^2$. This results in the equation

$$(y^2 + p + z)^2 = (p + 2z)y^2 - qy + (p^2 - r + 2pz + z^2). \tag{4.17}$$

The trick is now to choose z so as to make the right-hand side in equation 4.17, a quadratic in y, also a perfect square. This requires the discriminant of the quadratic to vanish, i.e.,

$$4(p + 2z)(p - r + 2pz + z) = q^2. \tag{4.18}$$

Thus we arrive at a cubic equation for z (called the *resolvent cubic*), which can be solved by the methods of the previous section. Once this has been done, square roots can be extracted in equation 4.17. This produces a pair of quadratic equations in y that can then be solved.

We illustrate the procedure with an example from *Ars Magna.* Cardano considered the equation (Chapter 39, Problem 9).

$$y^4 - 10y^2 + 4y + 8 = 0.$$

Following the initial steps above leads to equation 4.18 in the form

$$4(2z - 10)(92 - 20z + z^2) = 16,$$

which after simplification, becomes

$$z^3 - 25z^2 + 192z = 462.$$

This equation has as a root $z = 7$. With this value of z, equation 4.17 becomes

$$(y^2 - 3)^2 = 4y^2 - 4y + 1 = (2y - 1)^2. \tag{4.19}$$

Taking the square root results in the quadratic equations

$$y^2 - 3 = 2y - 1$$

and

$$y^2 - 3 = 1 - 2y.$$

Solving these equations, we have the four roots

$$y = 1 + \sqrt{3},\ 1 - \sqrt{3},\ -1 + \sqrt{5},\ -1 - \sqrt{5}.$$

4.6 DE MOIVRE'S THEOREM

Abraham de Moivre (1667–1754) was a French mathematician who made contributions in many areas of mathematics. He was born in Vitry-le-François, but moved to England at a young age to avoid persecution of the Huguenots in France. There he made the acquaintance of Newton, Halley, Stirling, and other noted mathematicians, who influenced the course of his scientific work. His contributions include the discovery of the closed form expression for the Fibonacci numbers (Binet's formula), a multinomial version of Newton's binomial theorem, introducing the central limit theorem into probability (in which subject he wrote an influential book titled, *The Doctrine of Chances*), and a preliminary form of Sterling's formula. (We will discuss these developments in later chapters). de Moivre's most well-known result is the following theorem, which dates to 1722.

Theorem 4.4. *(de Moivre) For any angle θ and positive integer n,*

$$(\cos\theta + i\sin\theta)^n = \cos n\theta + i\sin n\theta. \tag{4.20}$$

An expression $E(\theta)$ in θ with the property $\big(E(\theta)\big)^n = E(n\theta)$. It suggests, does it not, that $E(\theta)$ is some sort of exponential function, $E(\theta) = A^\theta$. We will see later that this is indeed the case, with $A = e^i$, where $e = 2.71828...$ is the exponential number and $i = \sqrt{-1}$. However, this result will need to await some further developments. (What does it even mean to raise a real number to an imaginary power?)

The proof of Theorem 4.4 that we are about to give makes use of a method known as *induction*. The idea is as follows. Suppose one wishes to prove that a certain assertion about the integers, call it $P(n)$, holds for every integer $n = 1, 2, 3, ...$ One way to do this is as follows. First, check that $P(1)$ holds. Then, *assuming* $P(n)$ holds, prove that $P(n+1)$ holds (this is called the inductive step). Then we have the string of implications

$$P(1) \implies P(2) \implies P(3) \implies \dots,$$

thereby showing that $P(n)$ holds for all positive integers n.

FIGURE 4.9 Abraham de Moivre

As an example, we use induction to prove the statement

$$P(n): \quad 1+2+3+\cdots+n=\frac{n(n+1)}{2}. \tag{4.21}$$

As is customary, we write the sum in (4.21) (and later sums) in the more concise summation notation

$$\sum_{k=1}^{n} k = 1+2+3+\cdots+n.$$

First note that $P(1)$ is clearly true.

Assuming $P(n)$, we have

$$\begin{aligned}\sum_{k=1}^{n+1} k &= \sum_{k=1}^{n} k + (n+1)\\ &= \frac{n(n+1)}{2} + (n+1)\\ &= \frac{n^2+3n+2}{2} = \frac{(n+1)(n+2)}{2}\\ &= \frac{(n+1)([n+1]+1)}{2}.\end{aligned}$$

Thus $P(n+1)$ is established, and (1.2) follows for all n, by induction.

Exercise 4.6. *Prove by induction*

$$\sum_{k=1}^{n} k^2 = \frac{n(n+1)(2n+1)}{6}.$$

We now prove de Moivre's theorem. Denote the assertion in the theorem by $P(n)$ and note that $P(1)$ is obviously true. Now, assume $P(n)$. Then

$$(\cos\theta + i\sin\theta)^{n+1} = (\cos\theta + i\sin\theta)^n(\cos\theta + i\sin\theta)$$

$$= (\cos n\theta\cos\theta - \sin n\theta\sin\theta) + (\sin n\theta\cos\theta + \cos n\theta\sin\theta)i$$

$$= \cos(n+1)\theta + i\sin(n+1)\theta,$$

according to the trigonometric identities

$$\sin(A+B) = \sin A\cos B + \cos A\sin B,$$

$$\cos(A+B) = \cos A\cos B - \sin A\sin B.$$

Thus the inductive step is complete and with it, the proof of de Moivre's theorem.

The Fundamental Theorem of Algebra states that every polynomial equation has precisely as many complex roots as its degree. The following immediate consequence of de Moivre's theorem provides an explicit representation of these roots for the equation $z^n = 1$.

Theorem 4.5. *The nth roots of the complex number $z = a + bi$ are given by*

$$\sqrt[n]{z} = \sqrt[n]{r}\Big(\cos\big(\frac{\theta + 2k\pi}{n}\big) + i\sin\big(\frac{\theta + 2k\pi}{n}\big)\Big),\; k = 1, 2, \ldots, n, \qquad (4.22)$$

where $z = r(\cos\theta + i\sin\theta)$ denotes the polar form of z (see figure 4.10).

Exercise 4.7* *Use Theorem 4.5 to compute the four values of $\sqrt[4]{1+i}$.*

4.7 LAGRANGE AND VANDERMONDE

After the work of Ferrari, the study of polynomial equations naturally turned to those of degree five and higher, but little essential progress was made in this area for at least 200 years. In 1770, Lagrange published a large and highly influential treatise on polynomial equations, titled *Réflexions sur la résolution algēbrique des équations.* Here Lagrange unified the preceding methods of solution of the quadratic, cubic, and quartic equations and devised a general approach for the solution of polynomial equations. We see a shift in emphasis away from the clever transformations and manipulations of Tschirnhaus, Tartaglia, del-Ferro, Ferrari, etc. toward theory, which paved the way for the

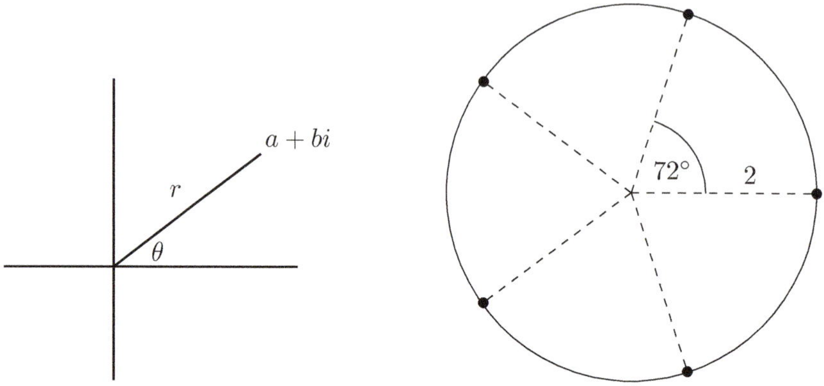

FIGURE 4.10 Polar coordinates of a complex number and the fifth roots of 32 in the complex plane

FIGURE 4.11 Joseph-Louis Lagrange

later advances of Ruffini, Lagrange, Abel, and Galois and which is characteristic of much of modern mathematics.

Lagrange's memoir starts with a review of the methods of his predecessors, but does much more than that. His objective is to explain not just *how* these methods work, but *why*. The previous methods all share a common feature, namely to reduce the solution of the proposed equation to an *auxiliary* equation of lower degree. Lagrange conceived the highly original idea to reverse the flow of information and express the roots of the auxiliary equations as functions of the roots of the proposed equations. This approach sheds new light on the methods.

In pursuing this line of investigation, Lagrange singled out the following combination of the roots $x_1, x_2, \ldots, x_n$ of the polynomial

$$t(\omega) = x_1 + \omega x_2 + \omega^2 x_3 + \cdots + \omega^{n-1} x_n, \tag{4.23}$$

where ω is an nth root of unity, not equal to 1, i.e., satisfies

$$\omega^{n-1} + \omega^{n-2} + \ldots \omega^{n-3} + \cdots + 1 = 0.$$

These polynomials are now called *Lagrange resolvents*, although they actually originated in works of Euler and Bézout.

Lagrange established the inversion formula

$$x_i = \frac{1}{n}\Big(\sum_{\omega} \omega^{-(i-1)} t(\omega)\Big), \tag{4.24}$$

where the sum runs over all the nth roots of unity.

Lagrange devised a unified plan of attack on equations of degree two, three and four. However, he found that when he applied the same method to equations of degree five and higher, a mysterious and disconcerting pattern appeared. Instead of resulting in an equation of lower degree, the argument gave rise to one of *higher* degree. For example, when applied to the quintic equation, Lagrange's method produces an equation of the sixth degree, then one of the tenth degree, and the situation gets worse and worse. Thus the theory that Lagrange had developed would not resolve the issue of solvability of polynomials of degree five and higher.

Notwithstanding, Lagrange had made a major step forward in the solvability of polynomial equations. In particular, he introduced roots of unity into the problem, established the resolvents (4.23) as a key tool, and carried out the type of intricate calculations with permutations that would play a decisive role in the resolution of the problem.

In Lagrange's own words: *It would be opportune to apply it* (his theory) *to the equations of the fifth and higher degrees, whose solution is so far unknown; but this application requires a too large amount of researches and combinations, whose success is, for that matter, still very dubious, for us to tackle this problem now; we hope, however to come back to it at another time, and we will be content to have here set the foundations of a theory which seems to us new and general.*

The Work of Vandermonde

Almost simultaneously with the publication of Lagrange's memoir, another treatise on polynomial equations appeared with a very similar name, *Mémoir sur la résolution des équations* by the French mathematician Alexandre Théophile Vandermonde.

Vandermonde was a mathematician of much lesser renown than Lagrange. Ironically, his name is known today largely for a determinant that does not even appear in his works. Nonetheless, Vandermonde made a startling algebraic breakthrough which had eluded Lagrange (and everyone else up to that time), namely the derivation of radical expressions for the 11th roots of unity.

Vandermonde's paper consisted of two parts. In the first part, he studied general polynomial equations and obtained many of the results in Lagrange's memoir. Vandermonde also worked with resolvents, which he defined by

$$V_i = \rho_1^i x_1 + \rho_2^i x_2 + \cdots + \rho_n^i x_n,$$

where $\rho_1, \rho_2, \ldots, \rho_n$ are the nth roots of unity, including 1. (Note the similarity with Lagrange's resolvents). Vandermonde derived the formula

$$x_i = \frac{1}{n}\left(\sum_{1=1}^{n} x_i + \sum_{i=1}^{n-1} \sqrt[n]{V_i^n}\right), \tag{4.25}$$

where the particular root x_i obtained depends on which nth roots are extracted inside the second summation. Thus Vandermonde's approach *exploited,* in a sense, the ambiguity inherent in extracting roots.

The second part of the paper is concerned with the study of cyclotomic equations. Vandermonde makes a detailed study of the equation

$$x^{10} + x^9 + x^8 + \cdots + x + 1 = 0. \tag{4.26}$$

de Moivre had previously observed that the substitution $y = x + x^{-1}$ reduces (4.26) to the lower degree equation

$$y^5 + y^4 - 4y^3 - 3y^2 + 3y + 1 = 0$$

but was unable to solve this equation. Vandermonde make use of the slightly different substitution

$$z = -(x + x^{-1}),$$

which yields the equation

$$z^5 - z^4 - 4z^3 + 3z^2 + 3z - 1 = 0. \tag{4.27}$$

The five roots of this equation can be expressed trigonometrically as

$$a = -2\cos\frac{2\pi}{11},\ b = 2\cos\frac{4\pi}{11},\ c = -2\cos\frac{6\pi}{11},\ d = -2\cos\frac{8\pi}{11},\ e = -2\cos\frac{10\pi}{11}.$$

Exercise 4.8. *Use de Moivre's theorem to prove this.*

In general, there is no formula to solve a fifth degree equation by radicals. However, (4.27) is no common or garden quintic; its roots are cosines of angles in arithmetic progression (all multiples of $2\pi/11$). Vandermonde notes that the trigonometric identity

$$2\cos\alpha\cos\beta = \cos(\alpha+\beta) + \cos(\alpha-\beta)$$

implies algebraic relations among the roots. For example, taking $\alpha = \beta = \frac{2\pi}{11}$ yields

$$a^2 = -b + 2$$

while choosing $\alpha = \frac{2\pi}{11}, \beta = \frac{4\pi}{11}$, we have

$$ab = -a - c.$$

The full list of relations thus obtained is:

$$\begin{array}{llll} a^2 = -b+2, & b^2 = -d+2, & c^2 = -e+2, & d^2 = -c+2, e^2 = -a+2, \\ ab = -a-c, & bc = -a-e, & cd = -a-d, & de = -a-b, \\ ac = -b-d, & bd = -b-e, & ce = -b-c, & \\ ad = -c-e, & be = -c-d, & & \\ ae = -d-e. & & & \end{array}$$

The relevance of these relations to Vandermonde is that they can be used to reduce any polynomial in the roots a, b, c, d, e to a *linear* combination of the roots.

As an example, consider the polynomial $P = 3a^2c^2 - 4bc^2 + 5ae - 2e^2$, which we chose randomly. Using the relations several times, we have

$$\begin{aligned} P &= 3(-b+2)(-e+2) - 4(-a-e)c + 5(-d-e) - 2(-a+2) \\ &= 3(be - 2b - 2e + 4) + 4ac + 4ec - 5d - 5e + 2a - 4 \\ &= 3(-c-d) - 6b - 6e + 12 + 4(-b-d) + 4(-b-c) - 5d - 5e + 2a - 4 \\ &= 2a - 14b - 7c - 12d - 11e - 4. \end{aligned}$$

Vandermonde now makes a second, strikingly original, observation, which forms the crux of his argument: *there exists a cyclic*[1] *permutation that preserves the relations among the roots.* The permutation (denote it σ) is

$$a \mapsto b \mapsto d \mapsto c \mapsto e \mapsto a.$$

By this is meant the following: choose any one of the relations, say $cd = -a-d$. Replace the terms by their associates under the permutation, so a by $\sigma(a) = b$,

[1] Meaning, as the premutation is applied repeatedly, each of the roots gets taken to every other root exactly once before returning to itself.

c by $\sigma(c) = e$, and d by $\sigma(d) = c$. This gives $ec = -b - c$, and this is again one of the relations!

Vandermonde puts these observations to work as follows. Let $\omega \neq 1$ denote any one of the complex fifth roots of 1, e.g.,

$$\omega = \frac{\sqrt{5}-1}{4} + \Big(\frac{3-\sqrt{5}}{8}\Big)i$$

(see Section 1.6 where this is derived.) Vandermonde defines his resolvents V_i by

$$V_i(a,b,c,d,e) = a + \omega^i b + \omega^{3i} c + \omega^{2i} d + \omega^{4i} e,\ 1 = 1,\ldots,5.$$

(The ordering of the powers of ω here is dictated by the permutation σ.) In view of the preceding discussion, there are linear expressions of the form

$$V_i(a,b,c,d,e)^5 = Aa + Bb + Cc + Dd + Ee + F, \tag{4.28}$$

with $A,\ldots,F$ explicitly computable (along the lines of the previous example) integer combinations of ω and its powers[2].

He now applies the permutation σ four times in (4.28). The point is that the coefficients $A,\ldots,F$ (which depend only upon the relations between the roots) *do not change.* The result is

$$V_i(b,d,e,c,a)^5 = Ab + Bd + Ce + Dc + Ea + F, \tag{4.29}$$

$$V_i(d,c,a,e,b)^5 = Ad + Bc + Ca + De + Eb + F, \tag{4.30}$$

$$V_i(c,e,b,a,d)^5 = Ac + Be + Cb + Da + Ed + F, \tag{4.31}$$

$$V_i(e,a,d,b,c)^5 = Ae + Ba + Cd + Db + Ec + F. \tag{4.32}$$

We also have

$$a + b + c + d + e = 1$$

since the sum of the roots is the negative of the coefficient of z^4 in equation 4.27. Furthermore, it is easy to check that the left-hand sides in each of the equations 4.28–4.32 is the same. This happens because various V_i's in the above equations differ from each other only by multiplication by an integer power of ω, and $\omega^5 = 1$. (This was the rationale for ordering the powers of ω in the definition of the original V_i as we did.)

Summing equations 4.28–4.32 yields

$$5V_i(a,b,c,d,e)^5 = A + B + C + D + E + 5F.$$

Vandermonde's concludes from (4.25), that

$$a,b,c,d,e = \frac{1}{5}\Big(1 + \sum_{i=1}^{4} \sqrt[5]{(A+B+C+D)/5+F}\Big).$$

---

[2]The coefficients $A,\ldots,F$ depend on i, but we have suppressed this in the notation.

The result is the formula

$$\cos\frac{2\pi}{11} = \frac{1}{5}\Bigg[\sqrt[5]{\frac{11}{4}\left(89+25\sqrt{5}-5\sqrt{-5+2\sqrt{5}}+45\sqrt{-5-2\sqrt{5}}\right)}$$
$$+\sqrt[5]{\frac{11}{4}\left(89+25\sqrt{5}+5\sqrt{-5+2\sqrt{5}}-45\sqrt{-5-2\sqrt{5}}\right)}$$
$$+\sqrt[5]{\frac{11}{4}\left(89-25\sqrt{5}-5\sqrt{-5+2\sqrt{5}}-45\sqrt{-5-2\sqrt{5}}\right)}$$
$$+\sqrt[5]{\frac{11}{4}\left(89-25\sqrt{5}+5\sqrt{-5+2\sqrt{5}}+45\sqrt{-5-2\sqrt{5}}\right)}\Bigg].$$

Exercise 4.9.* *Prove the claim that the expressions V_i defined in equations 4.28–4.32 differ only by multiplication by integer powers of ω.*

Vandermonde claimed that this derivation is part of a general method which applies to cyclotomic equations of arbitrary degree. A comprehensive theory along these lines was later worked out by Gauss.

This is math magic of the highest order! Vandermonde had not only anticipated Gauss' theory of cyclotomy, but also hit upon the insight that would become the germ of Galois theory: in order to understand the solvability, or otherwise, of a given polynomial equation, it is necessary to analyze the group of permutations which *preserve the relations between the roots*. We describe something of these developments in the next two sections.

4.8 GAUSS AND THE HEPTADECAGON

Carl Friedrich Gauss (1777–1855), born in Brunswick, Germany, is considered one of the greatest mathematicians, notable for his contributions to number theory, differential geometry, astronomy, mathematical physics, and much else. His name appears many times in this book.

The image of the "Prince of Mathematics" born of humble intellectual stock, is the stuff of lore. His father, although educated, held a series of menial jobs, while his mother was almost illiterate. Gauss' precocious mathematical talent is likewise legendary; how, at the age of seven and in response to a schoolmaster's assigned chore, he was able to immediately give the sum of the numbers 1 to 100. The young Gauss had realized that if one writes down the sum in the reverse order and adds the two equalities

$$S = 1 \quad +2 \ +3 \ +\cdots+99+100,$$
$$S = 100+99+98+\cdots+2 \ +1,$$

then the result is

$$2S = 101+101+\cdots+101 = 101\times 100 = 10100,$$

whence $S = 5050$.

Quickly recognizing his prodigious talent, Gauss' teachers brought him to the attention of the Duke of Brunswick who financed his further education. The Duke's stipend stopped after he was killed in battle, but by then Gauss was a world famous mathematician.

Gauss married twice and had six children in all. A victim of a domineering father, he became one himself. He forbade his sons from becoming mathematicians on the grounds that since their contributions would never measure up to his own, they would degrade the family name. In 1806, Gauss was appointed as director of the astronomical observatory at Göttingen, where he remained, proud and aloof, for the rest of his career. He died in 1855.

The problem of constructing regular polygons dates back to antiquity. As we have seen in Chapter 1, it was known in Euclid's time how to construct, with straightedge and compass, the equilateral triangle and the regular pentagon. No further progress in this direction was made until 1796, when Gauss, at the age of 19, solved the problem for the regular heptadecagon (17-sided polygon). This discovery, which is said to have been decisive to Gauss' decision to pursue a career in mathematics, was published in his monumental treatise on number theory, *Disquisitiones Arithmaticae* in 1801.

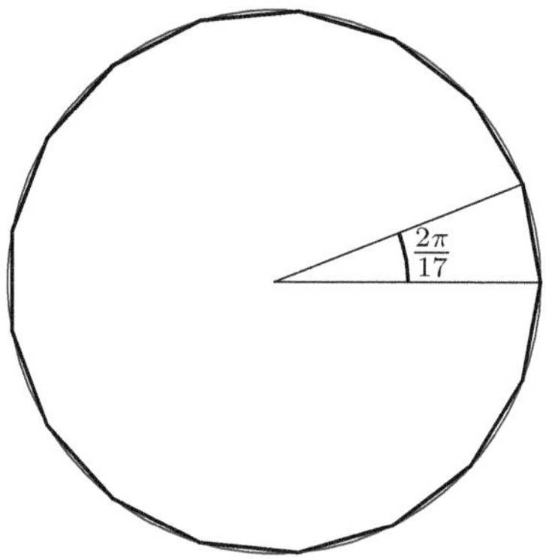

FIGURE 4.12 The heptadecagon

According to Theorem 1.4, constructibility of the heptadecagon follows from the existence of an arithmetic expression of the quantity $\cos\frac{2\pi}{17}$ (equivalently $\sin\frac{2\pi}{17}$) in terms of the integers and (iterated) square roots. Gauss obtained the expression

$$\cos\frac{2\pi}{17} = \frac{1}{16}\Bigg(-1+\sqrt{17}+\sqrt{34-2\sqrt{17}} \\ + \sqrt{68+12\sqrt{17}-16\sqrt{34+2\sqrt{17}}-2(1-\sqrt{17})\sqrt{34-2\sqrt{17}}}\Bigg). \tag{4.33}$$

FIGURE 4.13 Portrait of Carl Friedrich Gauss by Christian Albrecht Jensen, 1840

In order to see the gist of Gauss' derivation of formula (4.33) let us first describe his method in a much simpler setting, namely for the angle $2\pi/5$.

Denote

$$r = \cos\frac{2\pi}{5} + \sin\frac{2\pi}{5}i.$$

By de Moivre's theorem, $r^5 = \cos 2\pi + i\sin 2\pi = 1$. Thus r satisfies the equation

$$0 = r^5 - 1 = (r-1)(r^4 + r^3 + r^2 + r + 1).$$

Since $r \neq 1$, this implies

$$r^4 + r^3 + r^2 + r + 1 = 0. \tag{4.34}$$

The objective of Gauss' program is to solve equations of this type by reducing them to a series of nested quadratic equations (only one is necessary in this case).

Set $x_1 = r + r^4$ and $x_2 = r^2 + r^3$, so that, by (4.34),

$$x_1 + x_2 = -1. \tag{4.35}$$

We have

$$x_1 x_2 = r^3 + r^4 + r^6 + r^7.$$

Since $r^5 = 1$, we may replace r^6 by r and r^7 by r^2. Thus (4.34) implies

$$x_1 x_2 = r^3 + r^4 + r + r^2 = -1. \tag{4.36}$$

Solving for x_2 in (4.35) and substituting into (4.36) yields the quadratic equation

$$x_1^2 + x_1 - 1 = 0$$

from which we obtain

$$x_1 = \frac{\sqrt{5} - 1}{2}.$$

de Moivre's theorem implies

$$\begin{aligned} x_1 &= \left(\cos\frac{2\pi}{5} + \cos\frac{8\pi}{5}\right) + \left(\sin\frac{2\pi}{5} + \sin\frac{8\pi}{5}\right)i \\ &= \left(\cos\frac{2\pi}{5} + \cos\left(-\frac{2\pi}{5}\right)\right) + \left(\sin\frac{2\pi}{5} + \sin -\left(\frac{2\pi}{5}\right)\right)i \\ &= 2\cos\frac{2\pi}{5}. \end{aligned}$$

Hence

$$\cos\frac{2\pi}{5} = \frac{\sqrt{5} - 1}{4}.$$

It is important to note that this argument *will not work* with different choices of x_1 and x_2, e.g., $x_1 = r + r^2, x_2 = r^3 + r^4$. Then, there is no simple expression for $x_1 x_2$.

We now turn to Gauss' derivation of (4.33). Let

$$r = \cos\frac{2\pi}{17} + \sin\frac{2\pi}{17}i.$$

Arguing as above, we have

$$r + r^2 + r^3 + \ldots r^{16} = -1. \tag{4.37}$$

The brilliant idea at the heart of Gauss' argument is to arrange the powers of r from 1 to 16 in the following list:

$$r^3, r^9, r^{10}, r^{13}, r^5, r^{15}, r^{11}, r^{16}, r^{14}, r^8, r^7, r^4, r^{12}, r^2, r^6, r. \tag{4.38}$$

There is method in this madness; the powers of r in the ordering are the integers $3^k, k = 1, 2, \ldots$, modulo 17. (Gauss explores this issue in Chapter VII of *Disquisitiones* and provides a general treatment of the problem.) The list will be used to select terms for various expressions that will be introduced shortly.

Gauss defines two quantities x_1 and x_2 which he calls *periods* of length eight, by summing the alternate items in the list (4.38) starting with the second, and then with the first term; thus

$$x_1 = r^9 + r^{13} + r^{15} + r^{16} + r^8 + r^4 + r^2 + r,$$

$$x_2 = r^3 + r^{10} + r^5 + r^{11} + r^{14} + r^7 + r^{12} + r^6.$$

By (4.37)

$$x_1 + x_2 = -1. \tag{4.39}$$

Now comes the amazing part of the argument. Brute force calculation (if nothing else) of the 64 terms in the product x_1x_2 shows that each of the powers of r from 1 to 16 occurs exactly four times! Thus (4.37) implies

$$x_1x_2 = -4. \tag{4.40}$$

Solving (4.38) and (4.40) for x_1 and x_2 and noting that $x_1 > x_2$, yields

$$x_1 = \frac{1+\sqrt{17}}{2},$$

$$x_2 = \frac{1-\sqrt{17}}{2}.$$

Gauss then uses the even and odd scheme to define periods of length 4:

$$y_1 = r^{13} + r^{16} + r^4 + r,$$

$$y_2 = r^9 + r^{15} + r^8 + r^2,$$

$$y_3 = r^{10} + r^{11} + r^7 + r^6,$$

$$y_4 = r^3 + r^5 + r^{14} + r^{12}.$$

Then

$$y_1 + y_2 = x_1, \; y_3 + y_4 = x_2$$

with $y_1 > y_2$ and $y_3 > y_4$. Also,

$$y_1y_2 = y_3y_4 = -1.$$

Solving for y_1, y_2, y_3, y_4, we have

$$y_1, y_2 = \frac{x_1 \pm \sqrt{x_1^2+4}}{2},$$

$$y_3, y_4 = \frac{x_2 \pm \sqrt{x_2^2+4}}{2}.$$

Now come the final steps. Reducing powers modulo 17, gives

$$y_1 = r^{-4} + r^{-1} + r + r^4,$$

$$y_4 = r^{-5} + r^{-3} + r^3 + r^5.$$

Define $z_1 = r^{-1} + r, \;\; z_2 = r^{-4} + r^4$. Then

$$z_1 + z_2 = y_1$$

$$z_1 z_2 = y_4$$

with $z_1 > z_2$. Solving for z_1 gives

$$z_1 = \frac{y_1 + \sqrt{y_1^2 - 4y_4}}{2}.$$

Similarly to before,

$$\cos\frac{2\pi}{17} = \frac{1}{2}z_1 = \frac{y_1 + \sqrt{y_1^2 - 4y_4}}{4}. \tag{4.41}$$

Substituting for y_1 and y_4 in (4.41), then x_1 and x_2, and simplifying, yields (4.33).

Exercise 4.10. *Show that $x_1 > x_2$ in Gauss' proof of (4.33).*

A literal ruler and compass construction of the heptadecagon based on Gauss' formula (4.33) was devised by Richmond in 1893. An animation of Richmond's construction can be found online at https://www.youtube.com/watch?v=xGUWVPOks00.

Which Other Regular Polygons Are Constructible?

Let P_m denote the regular polygon with m sides. In Section VII of *Disquisitiones*, Gauss extended his work to prove that P_m is constructible if m is a prime of the form $2^k + 1$, for k a positive integer. Prime numbers of this form are named for Fermat, who studied them prior to Gauss.

Gauss claimed that these are the only primes for which P_m is constructible and this claim was later proved by Wantzel in 1837.

Since an arbitrary angle can be bisected with straightedge and compass, it follows that $P_{2m}, P_{4m}, P_{8m}, \ldots$ are constructible whenever P_m is constructible. Combining these results leads to the conclusion that a regular polygon with N sides is constructible if and only if

$$N = 2^k F_1 F_2 \ldots F_n,$$

where k is a positive integer and $F_1, F_2, \ldots, F_n$ are distinct Fermat primes. The only such primes known to date are 3, 5, 17, 257, and 65537.

Exercise 4.11. *Prove that if $2^n + 1$ is a Fermat prime, then $n = 2^k$ for a positive integer k.*

4.9 UNSOLVABILITY OF THE QUINTIC AND HIGHER DEGREE EQUATIONS

Following the unsuccessful attempts by Euler, Tschirnhaus, Bézout, Vandermonde, Lagrange and a host of others to solve the quintic equation by radicals (i.e., expressions involving only the arithmetic operations and extractions of roots), a feeling began to emerge that perhaps such a solution did not exist. Gauss suggested as much in *Disquisitiones*, but did not venture anything in the way of a proof.

In 1799, the Italian mathematician Paolo Ruffini published a book, *Teoria Generale delle Equazioni* which purported to prove that the general quintic equation cannot be solved by radicals. The proof was found to be incomprehensible and was regarded with suspicion.[3] Later Ruffini published several simplified versions of his proof, but these too were not generally accepted. In retrospect, it seems that Ruffini's work did, in fact, contain a significant gap. Nonetheless, he had made significant inroads into the problem. In 1824, the brilliant Norwegian mathematician Neils Henrik Abel (1802–1829) published what is now regarded as the first definitive proof of the insolvability of the quintic equation by radicals. The result is known today as the *Abel–Ruffini theorem.*

FIGURE 4.14 Neils Henrik Abel

At this point, two classes of polynomial equations had been treated, with totally opposite results. On the one hand, there was the Abel–Ruffini theorem.

[3]A notable exception is Cauchy, who regarded Ruffini's work as correct and was very supportive of it.

FIGURE 4.15 Évariste Galois

Yet Gauss had proved that cyclotomic equations

$$x^n + x^{n-1} + x^{n-2} + \cdots + 1 = 0$$

are solvable (in radicals).[4] The question remained, precisely which polynomial equations are solvable?

In 1832, a young Frenchman named Évariste Galois was killed in a duel. Galois was born in a village near Paris in October of 1811. His father was at one time the village mayor and his mother a homemaker with a love of learning. Galois showed great mathematical ability from an early age, but was denied admission to the prestigious École Polytechnique (supposedly) for refusing to answer the examiner's questions. Galois was further frustrated and embittered when a memoir outlining his extraordinary ideas on algebraic equations was rejected by the French Academy as being incomprehensible. In retrospect, the rejection seems perfectly reasonable; Galois had not taken the trouble to sufficiently explain his work.

In July of 1829, Galois' father committed suicide following a political scandal and thereafter Galois' behavior became more and more erratic. In April of 1831, he was arrested for allegedly making threats against the king and thrown

[4] In fact, Gauss' proof of this result also had a gap, but this was filled in by later mathematicians.

in jail. The fatal duel, which took place on May 30, 1832 was ostensibly over a woman, but may well have been trumped up by Galois' political enemies as a way of removing him from the scene. Presumably having a premonition of the outcome, Galois spent the night before the duel setting down his mathematical thoughts in a letter to a friend. Galois was shot in the stomach and died of his wounds the following day. He was twenty years of age.

A decade later, Joseph Liouville would address the French Academy of Sciences with the words:

I hope to interest the Academy in announcing that among the papers of Évariste Galois I have found a solution, as precise as it is profound, of this beautiful problem: whether or not there exists a solution (to polynomial equations) by radicals ...

The mathematical ideas which Galois set down in the ten page letter shortly before his death contain the essence of what is now known as *Galois theory.* Here Galois formulated a theory of polynomial equations which subsumes all the efforts of his predecessors and goes far beyond them. But the importance of Galois theory far exceeds even this. The subject has become an area of study in its own right and is a major tool in modern day research.

While it is not possible to give here anything like a detailed account of Galois theory, we will try to describe some of the underlying ideas. At the heart of the matter lie two mathematical objects known as fields and groups, so it is as well to start by discussing these.

An *(algebraic) field* is a set of numbers F which includes the rational numbers and is closed under the arithmetic operations of addition, subtraction, multiplication, and division, i.e., if $x, y \in F$, then $x + y, x - y, xy \in F$ and if $y \neq 0$, then $x/y \in F$.

Familiar examples of fields are the set of rational numbers $\mathbb{Q}$, the set of real numbers $\mathbb{R}$, and the set of complex numbers $\mathbb{C}$.

Another example is given in the following exercise.

Exercise 4.12. *Prove that the set* $G \equiv \{p + r\sqrt{3}/\ p, q \in \mathbb{Q}\}$ *is a field.*

The theory developed around fields is a powerful tool in its own right. This is illustrated by the following two problems that date back to antiquity. In 1837, Wantzel used field theory to prove the impossibility of both constructions.

Trisecting the angle: Using only straight edge and compass, trisect an arbitrary angle.

Doubling the cube: Using only straight edge and compass, construct a cube whose volume is exactly twice that of a given cube.

Important notions in field theory are those of a field extension and its degree. These are defined as follows.

Given a field F and elements $\alpha_1, \alpha_2, \ldots, \alpha_m \notin F$, we define the augmented field $F[\alpha_1, \alpha_2, \ldots, \alpha_m]$ to be the smallest field containing F and also $\alpha_1, \alpha_2, \ldots, \alpha_m$. For example, the field $\mathbb{Q}[\sqrt{2}, \sqrt{3}]$ is the set

$$\{p + q\sqrt{2} + s\sqrt{3} + t\sqrt{6}/\ p, q, r, s \in \mathbb{Q}\}.$$

Let F and G denote two fields with $F \subset G$. Suppose there exist elements $\alpha_1, \alpha_2, \ldots, \alpha_m \in G$ with the properties:

The only $r_1, r_2, \ldots, r_m \in F$ such that $r_1\alpha_1 + r_2\alpha_2 + \cdots + r_m\alpha_m = 0$ are $r_1 = r_2 = \ldots r_m = 0$. (We say the set $\{\alpha_1, \alpha_2, \ldots, \alpha_m\}$ is *linearly independent* over F).

Every element $g \in G$ is expressible as $g = r_1\alpha_1 + r_2\alpha_2 + \cdots + r_m\alpha_m$, for some $r_1, r_2, \ldots, r_m \in F$. (We say the set $\{\alpha_1, \alpha_2, \ldots, \alpha_m\}$ *spans* G).

Then we say that G is a *field extension* of F of *degree* m.[5] This is denoted

$$[F : G] = m.$$

We illustrate these definitions with the field G in Exercise 4.12. The spanning condition obviously holds for $\alpha_1 = 1$ and $\alpha_2 = \sqrt{3}$ by the definition of G.

To prove linear independence, suppose there exist $r_1, r_2 \in \mathbb{Q}$ such that

$$r_1(1) + r_2\sqrt{3} = 0.$$

Suppose at least one of $r_1, r_2 \neq 0$ (let it be r_2, the argument obviously works equally well if it is r_1.) Then we have

$$\sqrt{3} = \frac{-r_1}{r_2},$$

a contradiction, since $\sqrt{3}$ is irrational. Thus $r_1 = r_2 = 0$ and so

$$[\mathbb{Q}, G] = 2.$$

Wantzel's solution in the negative to the angle trisection problem follows the following lines. Let H denote an augmented field formed by adjoining to $\mathbb{Q}[i]$ (the lattice of rational points in $\mathbb{C}$), any finite set of points obtained from ruler and compass constructions. Because these arise from the intersections of lines and circles, and hence satisfy quadratic equations, it can be shown that

$$[\mathbb{Q}, H] = 2^n, \tag{4.42}$$

where n denotes the number of adjoined points.

[5]Those readers who have studied Linear Algebra might recognize these definitions as saying that the set $\{\alpha_1, \alpha_2, \ldots, \alpha_m\}$ forms a basis of G as a vector space over F, and the dimension of G is m.

Wantzel uses (4.42) to prove that it is impossible to trisect the angle $\pi/3$, or 60°. This is obviously equivalent to constructing the point $(x, 0)$ where $x = \cos(\pi/9)$. The trigonometric identity

$$\cos(3\alpha) = 4\cos^3(\alpha) - 3\cos(\alpha)$$

shows that x satisfies the cubic equation

$$8x^3 - 6x - 1 = 0. \tag{4.43}$$

Equation 4.43 has no rational roots, as can easily be checked by the rational roots theorem, an elementary result in algebra. It follows that the equation cannot be factored with rational coefficients. (It is said to be *irreducible* over $\mathbb{Q}$.) This can be shown to imply that the degree of any field extension of $\mathbb{Q}$ containing x is divisible by 3. This contradicts (4.42), proving the angle trisection is impossible.

A similar argument works to prove the impossibility of doubling the cube.

We now turn to the definition of a group.

A *group* in mathematics is a set G, together with an operation $\circ$ defined on G, satisfying the three properties:

Associativity: For any three elements a, b, c in G, we have $(a \circ b) \circ c = a \circ (b \circ c)$.

Existence of an identity element: There is $e \in G$ such that $a \circ e = e \circ a = a$ for all $a \in G$.

Existence of inverse elements: for all $a \in G$, there exists $b \in G$ such that $a \circ b = b \circ a = e$.

The groups in this discussion consist of permutations of the roots of polynomials. The *Galois group* of a polynomial

$$p(x) = a_n x^n + a_{n-1} x^{n-1} + \cdots + a_0$$

can be broadly defined as *the set of those permutations of the roots of p which preserve the algebraic relations among the roots* (if any such relations happen to exist).

Recall that this is precisely the device used by Vandermonde to solve the cyclotomic equation of degree 10, as described in Section 4.7, with the permutation of roots $\sigma : \ a \mapsto b \mapsto d \mapsto c \mapsto e \mapsto a$ playing a central role. The Galois group in this instance is

$$G = \{e = \sigma^0, \sigma, \sigma^2, \sigma^3, \sigma^4\}. \tag{4.44}$$

A group of this type, all of whose elements consist of powers of a single one of them, is called a *cyclic* group and is typical of the Galois group of a cyclotomic equation. At the other extreme, in the case of the general quintic equation (where it is assumed *a priori* that *no algebraic relations exist* among the roots), the Galois group is the set of all $5! = 120$ permutations of the roots. This group is called the *symmetric group on five letters* and is denoted S_5.

In spite of this rather simple sounding definition of the Galois group, it can be a tricky object to compute. When it comes to calculating the Galois group of a cyclotomic equation, we are in the privileged position of knowing at the outset the roots of the equation (albeit in trigonometric form), courtesy of de Moivre's theorem. This will not, of course, generally be the case. Without knowing the roots, how can it be ascertained whether or not algebraic relations exist among them and, if so, which permutations preserve these relations?[6] This would seem an insurmountable obstacle. Remarkably, it turns out that enough information can often be extracted from a given problem to make the Galois group a useful tool.

A solution in radicals to a polynomial equation is well expressed in the language of field extensions. Consider, for example, the cubic equation

$$x^3 - x + 3 = 0.$$

Cardano's formula yields the following expression for the roots

$$x = \sqrt[3]{-\frac{3}{2} + \sqrt{\frac{239}{27}}} + \sqrt[3]{-\frac{3}{2} - \sqrt{\frac{239}{27}}}.$$

Consider how this expression is built from the coefficients of the equation. The first step is to calculate the 239/27. (The introduction of a fraction does nothing to enlarge the field $\mathbb{Q}$ of the coefficients.) The square root is calculated. Then the two cube roots are extracted and summed to give the root(s) x of the equation. This creates a chain, or *tower* of field extensions leading from the field of rational numbers to the field containing the roots of the equation,

$$\mathbb{Q} \subset \mathbb{Q}\left[\sqrt{\frac{239}{27}}\right] \subset \mathbb{Q}\left[\sqrt{\frac{239}{27}}, \sqrt[3]{-\frac{3}{2} + \sqrt{\frac{239}{27}}}, \sqrt[3]{-\frac{3}{2} - \sqrt{\frac{239}{27}}}\right].$$

Note that each time the root of a number is extracted to produce a quantity not in an existing field, the field is extended.

The existence of a solution by radicals to any polynomial equation likewise implies a tower of field extensions from the field of rationals to the field K containing the roots:

$$\mathbb{Q} \subset K_1 \subset K_2 \subset ... \subset K_m = K. \tag{4.45}$$

Galois' fundamental theorem asserts that there exists a sequence of nested subgroups

$$G_0 = \{e\} \subset G_1 \subset G_2 \ldots, \subset G_m = G$$

[6] Galois himself points to this strange aspect of his work in the fateful letter.

(referred to as a *composition series*) of G which correspond precisely to the intermediate fields K_j, but in reverse order (G_0 corresponding to K_m, G_1 to K_{m-1}, etc.). Also, that the subgroups G_j are imbedded in each other in a specific way (this property is called *normality*).[7] When the Galois group of an equation admits a composition series of this type, it turns out that the equation is solvable by radicals. Otherwise, not. Furthermore, in the former case, analysis of the composition series reveals the solution to the equation.

It can be shown that cyclic groups of all orders admit such composition series, as do the symmetric groups S_2, S_3, S_4. Thus the corresponding equations are solvable. Conversely, one can be show that no such composition series exist for S_n, when $n \geq 5$. This provides an independent proof of the Abel–Ruffini theorem, that the general equation of degree five or more is unsolvable.

Abel and Ruffini had proved that there is no common formula with radicals (such as Cardano's) for solving polynomial equations of any given degree, when the degree exceeds four. The generality of this statement is also, in a sense, its weakness. It could nonetheless be the case that the equations all have solutions in radicals, but require a series of different formulas for their solution. Further evidence of the power of Galois theory is provided by the following theorem, which shows that this is not so.

Theorem 4.6. *Suppose p is a quintic polynomial with precisely two (non-real) complex roots, irreducible over the rationals. Then the Galois group of p is S_5, hence p cannot be solved with radicals.*

An example is the innocuous looking equation

$$x^5 - 6x + 3 = 0.$$

Exercise 4.13. *Verify that the polynomial in this equation satisfies the conditions of Theorem 4.6.*

At the risk of belaboring the point, not only is there no general formula in radicals for the solution of quintic equations, but the roots of this specific equation *cannot even be expressed* in this form. No more definitive answer to the question of the solvability of polynomial equations by radicals could be imagined.

It is clear that Galois did not prove all of these results. For example, his memoir to the French Academy contained but one theorem. It is the beautiful (but seemingly useless if one is looking for an external condition to determine whether or not a given equation is solvable by radicals) criterion:

In order that a polynomial equation of prime degree be solvable by radicals it is both necessary and sufficient that all of its roots be rational functions of any two of them.

[7]Groups that possess this structure are used nowadays outside of the context of polynomial equations. Owing to their origin, they are termed *solvable* groups.

Nonetheless, the ideas were all there in his work. Perhaps the most remarkable feature of Galois' work is that, in his day the concept of group barely even existed. Galois had, so to speak, to invent the wheel in order to invent the automobile. Galois, more than anyone, is responsible for initiating the modern age of algebra, based on mathematical structures rather than computations. This approach was to bear copious fruit in the invention of the field of commutative algebra and the theories of ideals and rings developed by Ernst Kummer (1810–1893), David Hilbert (1862–1943), and Emmy Noether (1882–1935), in the nineteenth and twentieth centuries.

Suggestions for Further Reading

Jeanne-Pierre Tignol. *Galois Theory of Algebraic Equations.* World Scientific, 2nd edition, 2016.

FIGURE 4.16 Amalie Emmy Noether

CHAPTER 5

Calculus

This chapter opens with biographical sketches of Newton and Leibniz, commonly regarded as the originators of calculus, then proceeds to a discussion of the subject itself. In Section 5.2, readers are introduced to the notion of "limit," the key idea on which calculus is based. The next two sections discuss the operations of differentiation, concerned with calculating the slopes of tangent lines and rates of change, and integration, calculating areas under curves. In Section 5.4 we present the Fundamental Theorem of Calculus. This theorem establishes an inverse relationship between the processes of differentiation and integration and is the main tool by which to calculate integrals. In Section 5.5, we discuss the representation of the functions sine, cosine, exponential and logarithm, as *power series* (polynomials with an infinite number of terms). These representations result in, among much else, Euler's famous formula connecting the exponential function with sine and cosine.

5.1 NEWTON AND LEIBNIZ

Issac Newton (1642–1726) was born in the English hamlet of Woolsthorpe in the county of Lincolnshire. Newton's father, also named Isaac, had died three months prior to his birth. Newton was born prematurely and was not expected to survive: according to his mother, Hannah Ayscough, he could have "fit inside a quart mug" at birth. When Newton was three years old his mother remarried and went to live with her new husband, the Reverend Barnabas Smith, leaving her son to be raised by his maternal grandmother.

In the year 1661, Newton was admitted as a student to Trinity College at the University of Cambridge, where he came under the tutelage of Isaac Barrow, the Lucasian Professor of Mathematics. Barrow fostered Newton's interest in mathematics. By all outward indications, Newton was an unexceptional student, privately he had embarked on a career of unheralded mathematical discovery.

DOI: 10.1201/9781003470915-5

FIGURE 5.1 Portrait of Issac Newton by Sir Godrey Kneller

In 1665, the university was temporarily closed as a precaution against the plague and Newton returned home to Woolsthorpe. Over the next two years Newton originated some of his great mathematical and scientific discoveries, a generalized version of the binomial theorem valid for fractional exponents, the development of calculus, his theory of light (later published as *Optiks*), and the law of universal gravitation. (The story about the role of the falling apple in the discovery of the latter is likely apocryphal.)

In April of 1666, Newton returned to Cambridge and was elected a Fellow of Trinity College. In 1669, Barrow resigned his chair in favor of Newton. Newton remained at Cambridge until 1696, when he moved to London to take up the post of Warden of the Royal Mint. He was elected a Fellow of the Royal Society in 1672 and served as its president from 1703 to 1727. He was knighted by Queen Anne in 1705.

In 1687, Newton published his *Philosophiae Naturalis Principia Mathematica* (Mathematical Principles of Natural Philosophy), commonly known as the *Principia*. This book is widely regarded as the greatest single work in the history of science. Herein, Newton formulated his three laws of motion and his law of universal gravitation. The latter states: *Any two material bodies attract each other with a force varying directly with the products of their masses and inversely with the squares of the distance between them.* In symbols,

$$F = \frac{GMm}{d^2},$$

where F is the force of attraction between the bodies, M and m are their masses and d the distance between them. The symbol G is a universal constant, the same for any two bodies, anywhere in the universe. As a crowning achievement of this part of the work, Newton proved that the inverse square law for the force impressed on a heavenly body is equivalent to the body moving in a conic section (ellipse, parabola, or hyperbola), with the force lying at one of the foci; behavior that had been previously observed by Kepler and described in his empirical laws of planetary motion.

The last of the three parts of the *Principia* has the grand, and entirely appropriate, title *De Systemate Mundi* ("System of the World"). Here Newton applies his mathematics to the solar system, describing the movements of planets, satellites, comets, and even terrestrial phenomena such as the shape and the density of the earth and the progression of the tides.

Using Newton's work, the astronomer Edmund Halley, who did much to bring about the publication of the *Principia*, calculated the orbit of the Great Comet of 1682, now named in his honor. Halley determined the comet's periodic orbit and its return every 75.5 years. This led to Halley correctly identifying the comet as the same one that had attracted the attention of Kepler in 1607 and was observed by Peter Apian in 1531. The correct prediction of the comet's return in 1759 was taken as confirmation of Newton's astronomical theory.

Gottfried Wilhelm Leibniz (1646–1716) was born in the university town of Leipzig in Germany. His father, a jurist and a professor of moral philosophy at the university, died when Gottfried was 6 years old. Leibniz immersed himself in the world of his father's books. A precocious child, he taught himself Latin when he was 8 years old and entered Leipzig University at the age of 15, where his work soon outstripped that of his contemporaries.

At Leipzig, Leibniz received a traditional education in religion and philosophy, together with some rudimentary mathematics. Descartes *Géométrie*, which he tried to study on his own, was too complicated for him at that time. Leibniz graduated in 1663, at the age of 17. He was given a teaching position in the Department of Philosophy at Leipzig on the basis of a thesis which he later expanded into a work, *Ars Combinatoria*, published in 1666. In this book, Leibniz undertook a study of combinations and permutations. Leibniz also proposed the extremely original idea of establishing a calculus of reasoning, as it were, which he called *characteristica universalis*, in which all scientific ideas would be generated from an "alphabet of human thoughts." After further legal studies in Leipzig and having being denied a doctoral degree in this subject on the spurious grounds that he was too young, Leibniz took up a post in the service of the archbishop-elector of Mainz in 1667. Here

FIGURE 5.2 Portrait of Gottfried Wilhelm Leibniz by Bernhard Christoph Francke

he was charged with reforming the legal statutes. Except for periods of travel abroad, the rest of Leibniz's career was spent in residence at the courts of Mainz and Hanover, where he remained until his death in 1716.

Leibniz spent the years from 1672 to 1676 in an extended visit to Paris where he met the great Dutch scientist Christian Huygens. Recognizing the young man's enormous aptitude for mathematics, Huygens undertook to tutor Leibniz in the subject and it is in this period that Leibniz' mathematical genius blossomed.

In 1673, Leibniz crossed the channel to England on a diplomatic mission for the elector of Mainz, where he made the acquaintance of his fellow countryman Henry Oldenburg, the secretary of the Royal Society. Oldenburg introduced Leibniz to many of the leading British scientists of the day including Pell, Collins, Boyle, and Hooke. On the basis of his mathematical contributions and his invention of a calculating machine, Leibniz was elected a member of the Society.

The Priority Dispute

The seeds of the controversy were sown by John Wallis. By 1693, Newton had written out three separate accounts of his approach to calculus, but never showed any inclination to publish. The first printed account of Newton's work appeared in Wallis' *Opera Mathematica* (1693). Wallis warned his Cambridge colleague, "Your notions of fluxions pass on the Continent with great applause, by the name of Leibniz' differentials ... I have endeavored to do you justice in that point." The "justice" that Wallis had done Newton was to state in the preface to Volume I of the *Opera* that Newton's method had been sent to Oldenburg in 1676 to be communicated to Leibniz, thereby suggesting that Leibniz had plagiarized Newton's work.

The first serious salvo in the battle between the two protagonists was fired by Nicolas Fatio de Duiller, a Swiss mathematician of no particular note. Fatio emigrated to London in 1687 and managed to ingratiate himself into Newton's inner circle. Fatio had felt slighted by something Leibniz had said (or rather, not said) about him. In 1699, Fatio published a tract in which he amplified the charge that Leibniz's ideas had been obtained from his (Fatio's) idol Newton. Nothing more happened for almost five years, until Newton's *Optiks* appeared in 1704. In the preface to this work Newton stated: *In a letter written to Mr. Leibniz in the year 1676, and published by Dr. Wallis, I mentioned a method by which I found some general theorems about squaring curvilinear figures ... And some years ago I lent out a manuscript containing such theorems ...*

Newton had evidently become convinced that Leibniz had gotten the key to his calculus from perusing *De Analysi*, one of Newton's unpublished manuscripts. This is almost certainly untrue; Leibniz had seen something of the manuscript on one of his visits to London but only in a very abridged form, and in any case had come up with his own ideas some months earlier. Leibniz countered with an anonymous review in the *Acta Eruditorm* criticizing Newton's work and stating that his version of calculus was his own invention, discovered independently of Newton.

The great dispute was now well under way. A charge of plagiarism against Leibniz was made in 1708 by another of Newton's champions, the Scottish mathematician John Keil, who stated:

All these laws follow from that very celebrated arithmetic of fluxions which, without any doubt, Dr. Newton invented first, as can readily be proved by anyone who reads the letters about it published by Wallis; yet the same arithmetic afterwards, under a changed name and method of notation, was published by Dr. Leibniz ...

The article was published in the *Philosophical Transactions* of the Royal Society in 1710. Leibniz, trustful fellow, demanded that a Commission be set up by the Royal Society to investigate the matter.

It must be said that while Leibniz was hardly blameless in all that ensued, Newton acted abominably. Newton was, at that time, the head of the Royal Society. The report of the Commission, published under the title *Commercium Epistolicum* and supposedly composed by "a panel of impartial judges," was essentially written by him alone. The verdict may be guessed. An equally slanted review of the document, also the handiwork of Newton, although palmed off under the name of Keil, subsequently appeared in the *Philosophical Transactions* in 1715.

Issac Newton was a man with issues. In his youth he had reportedly expressed a wish to "burn my stepfather and mother Smith" in their bed.[1] Newton had no close friends, only sycophantic devotees. With anybody even remotely close to his colossal genius, Newton's relations were icy, often hostile. In his tenure as Warden, and later Master, of the Royal Mint, a position granted him as a sinecure, Newton took full advantage of the opportunity to vent his wrath against the miserable bunch of petty criminals and forgers he encountered, many of whom he sent to the gallows. In later life, Newton would fondly recall having "broken Leibniz' heart." There may be some truth in this; Leibniz died alone, his funeral attended by a single mourner.

The reality was not quite so stark. Leibniz had, in fact, many admirers among the best mathematicians on the continent, and his work would be taken up and continued by them. A prime example is the development of the calculus of variations by Euler and Lagrange in the second half of the eighteenth century.

Leibniz never denied Newton's priority in the invention of calculus. Newton contended that "second inventors have no claims." History has rendered a different verdict, crediting Newton and Leibniz equally as co-inventors of calculus.

Their Contributions to Calculus

As we have seen, some of the essential ideas of calculus had existed since ancient times. A rudimentary form of integration was formulated by Eudoxus in his method of exhaustion and used by Archimedes in calculating the area of a parabolic segment. Later on Cavalieri, Torricelli, Fermat, Wallis, and Barrow carried out calculations tantamount to integration and differentiation in their work on quadrature (calculating areas under curves) and the construction of tangents. Thus calculus was "in the air" by the time Newton and Leibniz arrived on the scene. What these two men did was to forge these ideas into a coherent body of thought, discover formulas and rules of manipulation, and, most importantly, elucidate the fundamental inverse relationship between the operations of differentiation and integration.

The primary motivation for the integral calculus was the problem of calculating the area lying beneath the graph of a given function. Wallis had earlier constructed a table for what we could now term the integrals

$$\int_0^a (1 - x^2)^n dx,$$

for certain integer values of n. Whereas Wallis always considered the upper

[1] There has been much speculation about abandonment issues as the source of the rage which Newton carried within him throughout his life.

limit of integration a as a fixed quantity, Newton had the insight to treat a as an independent variable, thereby creating a functional form of Wallis' identities. These read as follows

$$\int_0^x (1-t^2)dt = x - \frac{x^3}{3}$$
$$\int_0^x (1-t^2)^2 dt = x - \frac{2x^3}{3} + \frac{x^5}{5}$$
$$\int_0^x (1-t^2)^3 dt = x - x^3 + \frac{3x^5}{5} - \frac{x^7}{7}$$
$$\int_0^x (1-t^2)^4 dt = x - \frac{4x^3}{3} + \frac{6x^5}{5} - \frac{4x^7}{7} + \frac{x^9}{9}.$$

Since the formulas for the differentiation of integer powers of x: $\frac{d}{dx}x^n = nx^{n-1}$ was known from the work of Barrow, Newton realized immediately that these formulas reveal, to take the third one as an example, that

$$\begin{aligned}\frac{d}{dx}\int_0^x (1-t^2)^3 dt &= \frac{d}{dx}\left(x - x^3 + \frac{3x^5}{5} - \frac{x^7}{7}\right)\\ &= 1 - 3x^2 + 3x^4 - x^6\\ &= (1-x^2)^3.\end{aligned}$$

From these, and other cases, Newton deduced the Fundamental Theorem of Calculus

$$\frac{d}{dx}\int_0^x f(t)dt = f(x).$$

Newton used the fundamental theorem to derive his generalized binomial theorem. This led him to conclude that the rule for differentiating integer powers holds also in the fractional case, i.e.,

$$\frac{d}{dx}x^{\frac{m}{n}} = \frac{m}{n}x^{\frac{m-n}{n}}.$$

Newton had no compact notation to denote integrals and derivatives, and referred to them as *fluents* and *fluxions*, respectively. The very terms (meaning "flow" and "change") reveal Newton's mode of thinking, which was geometric and very much rooted in the physical world.

By contrast, Leibniz's approach to calculus was *algebraic*, arithmetic even. Leibniz conceived of the derivative of a function, which he expressed in the notation dy/dx, as literally the quotient of the two infinitesimally small quantities dy and dx. This harks back to Barrow's differential triangle alluded to in Chapter 3. Leibniz called it the "characteristic triangle." The term *infinitesimal* was understood by Leibniz to mean a quantity which is smaller than any given positive number, and yet non-zero. It should be noted that this concept does not make sense from the viewpoint of standard arithmetic, which holds that if $0 \leq \epsilon < a$ for all $a > 0$, then necessarily $\epsilon = 0$. Interestingly, in the 1960s a theory of non-standard arithmetic emerged in which ideas from mathematical logic are used to give meaning to this idea, thereby placing the work of Leibniz and his followers on a wholly rigorous basis.

In a similar vein, Leibniz' integral, which he expressed in the now universally adopted notation

$$\int f(x)dx$$

was conceived as a sum of rectangles with infinitesimal base length dx. (The origin of the integral sign $\int$ is as an elongated version of the letter S, standing for "sum.")

In a nutshell, in Leibniz' system, differentiation is understood as a process of differences and integration as a process of sums. Viewed from this perspective, the fundamental theorem appears almost trivial. Since

$$d\Big(\int f(x)dx\Big) = f(x)dx$$

as the last term in the sum, it ought to be correct to divide this by dx and conclude

$$\frac{d}{dx}\Big(\int f(x)dx\Big) = \frac{d(\int f(x)dx)}{dx} = f(x).$$

Consider further the example of the *chain rule*, the rule for differentiating the composition of two functions, $y = f(g(x))$. The rule asserts that the derivative $y' = f'(g(x)g'(x)$ (see Theorem 5.3). In Leibniz' differential notation, writing $z = g(x)$, this becomes the seemingly trivial arithmetic statement

$$\frac{dy}{dx} = \frac{dy}{dz} \cdot \frac{dz}{dx}.$$

It has been said that calculus, particularly as expressed by Leibniz, brings problems that once would have required the power of an Archimedes, within the range of the average college student (the type of student that attends classes on a regular basis and turns in their homework!).

Newton and Leibniz both made use of infinitesimals in calculating derivatives. An example is the following extract from Newton's treatise *De Methodis Fluxionem*. Newton posed the problem: *The relation between the fluents being given, to find the relation between the fluxions, and conversely.* He goes on to give several examples, one of which we quote verbatim:

"Thus, let any equation $x^3 - ax^2 + axy - y^3 = 0$ and substitute $x + \dot{x}o$ for x, $y + \dot{y}o$ for y, there will arise

$$x^3 + 3x^2\dot{x}o + 3x\dot{x}o\dot{x}o + \dot{x}^3o^3 - ax^2 - 2ax\dot{x}o - a\dot{x}o\dot{x}o + axy + ax\dot{y}o \\ + ay\dot{x}o + a\dot{x}o\dot{y}o - y^3 - 3y^2\dot{y}o - 3y\dot{y}o\dot{y}o - \dot{y}^3o^3 = 0$$

Now, by supposition, $x^3 - ax^2 + axy - y^3 = 0$, which therefore being expunged and the remaining terms divided by o, there will remain

$$3x^2\dot{x} + 3x\dot{x}\dot{x}o + \dot{x}^3oo - 2ax\dot{x} - a\dot{x}\dot{x}o + ax\dot{y} \\ + ay\dot{x} + a\dot{x}\dot{y}o - 3y^2\dot{y} - 3y\dot{y}\dot{y}o - \dot{y}^3oo = 0$$

But whereas o is supposed to be infinitely little, that it may represent the moment of quantities, the terms that are multiplied by it will be nothing in respect to the rest; I therefore reject them and there remains

$$3x^2\dot{x} - 2ax\dot{x} + ax\dot{y} + ay\dot{x} - 3y^2\dot{y} = 0.\text{''}$$

The method here is, in fact, identical to that of Barrow.

Early calculations such as this were met by skepticism. There is good reason for this. Consider what we just did, divide by a quantity o, then set o to 0. This seems reminiscent of a "paradox" I saw as a kid. The paradox claims to prove that $2 = 1$ and goes like this: let $a = b \neq 0$. Then

$$a^2 = ab \implies a^2 - b^2 = ab - b^2.$$

Factor out $a - b$ in both sides of the equation. We have

$$(a - b)(a + b) = (a - b)b.$$

Now cancel out $a - b$ to get $a + b = b$. Since $a = b$, this says $2b = b$ and dividing by b, we arrive at the absurd conclusion $2 = 1$.

The problem is of course, in the step where we divide by $a - b$. Since it was assumed $a = b$, we are dividing by 0, which is not valid. The "argument" attempts to divert the reader's attention away from this small detail.

The most articulate criticism in the early days of calculus came from the clergyman and philosopher bishop George Berkeley (1685–1753) who wrote in 1734:

And what are these fluxions? The velocities of evanescent increments. And what are these evanescent increments? They are neither finite quantities, nor quantities infinitely small, nor yet nothing. May we not call them ghosts of departed quantities?

It may have been that Newton himself distrusted calculations with infinitesimals, or perhaps it was a desire to avoid controversy that led him to rework the proofs of the numerous mathematical propositions in the *Principia*, which he had originally discovered by means of calculus, into a (much more complicated) form based on Euclidean geometry. It is a tribute to the awesome power of Newton's mind that he was able to accomplish such a feat. In any event, it would take until well into the next century for mathematicians to establish a solid theoretical foundation for the calculus of Newton and Leibniz.

5.2 LIMITS AND CONTINUITY

As noted above, the notion of infinitesimals on which calculus was originally based, is problematic. In the eighteenth century, a theory of limits emerged, thanks to the efforts of Cauchy and others, which placed the subject on a secure mathematical footing.

The idea of limit is actually rather simple. We illustrate it with an example. Consider the function

$$y = \frac{2x^2 - 5x - 3}{x - 3}. \tag{5.1}$$

Note that the expression is defined for all values x with the exception of $x = 3$, where it assumes the nonsensical value $0/0$. We are interested in the behavior of y for values of x in the immediate vicinity of 3. Some values are listed in the following tables:

x	y
2.5	6
2.9	6.8
2.99	6.98
2.999	6.998

x	y
3.5	8
3.1	7.2
3.01	7.02
3.001	7.002

It will be observed that as x approaches ever closer to 3 from either side, the y-values approach 7. This behavior is confirmed by the graph of function (5.1), shown in figure 5.4. We say: *the limit of y as x approaches 3 is 7*, and write

$$\lim_{x \to 3} y = 7. \tag{5.2}$$

As it happens, limits can be determined algebraically, without reference to specific points or graphs. Note that the numerator in (5.1) can be factored as $(2x + 1)(x - 3)$. Canceling the common terms $(x - 3)$ in numerator and denominator allows us to express the function more simply, as

$$y = 2x + 1, \quad x \neq 3, \tag{5.3}$$

where the caveat $x \neq 3$ is necessary to remind us that y is undefined at $x = 3$. Taking the limit in (5.3) as $x \to 3$, yields the value 7.

At this point, the preceptive reader may "smell a rat." We remarked above that (5.3) does not define y for $x = 3$, yet we plugged $x = 3$ into (5.3)

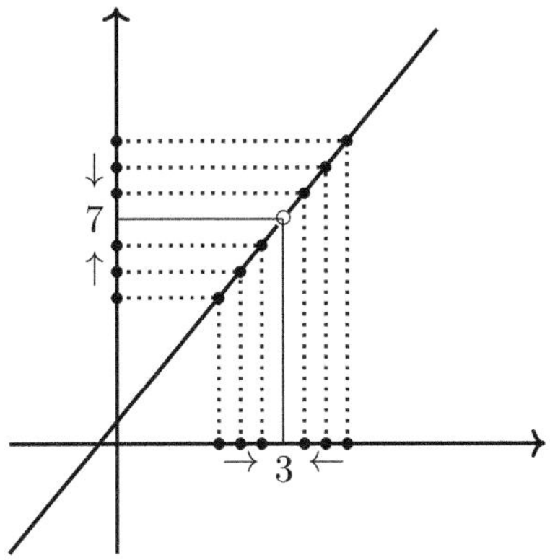

FIGURE 5.3

to calculate the limit. *Why is it valid to do this?* Because the limit is not concerned with the value of y at the point 3 itself, simply the behavior of y as x approaches 3, and for $x \neq 3$, the two expressions for y (the original and the reduced ones) agree.

Exercise 5.1.[⋆] *Compute the limit*

$$\lim_{x \to 9} \frac{\sqrt{x} - 3}{x - 9}.$$

(Hint: multiply the top and the bottom of the fraction by the conjugate of the numerator, $\sqrt{x} + 3$*).*

A Formal Definition of Limit

While the intuitive notion of limits introduced above suffices to make sense of the calculus of Newton and Leibniz, the rigorous development of the subject demands a more explicit approach to the concept. This is the subject of the present section.

When we write

$$\lim_{x \to a} f(x) = l \tag{5.4}$$

we are making the claim: *the function values* $y = f(x)$ *can be made arbitrarily close to* l *by making* x *sufficiently close to* a. The trick is to translate into mathematical language the terms "arbitrarily close" and "sufficiently close" and we do this with the following statement: *suppose an interval* J *centered at* l *is prescribed (of any width, no matter how small). Then a corresponding interval* I *centered at* a *exists such that if* x *lies in* I, *then* y *lies in* J. The intervals I and J are illustrated in figure 5.4.

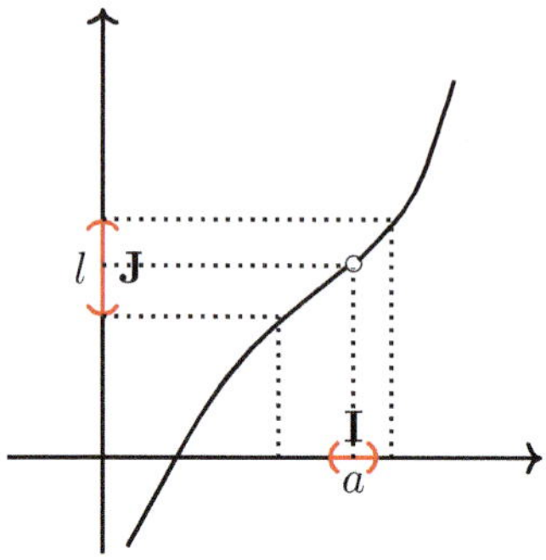

FIGURE 5.4

Denoting the widths of the intervals, as is customary, by the Greek letters ϵ and δ, we are arrive at the classic definition of limit: (5.4) is said to hold if, *given any* $\epsilon > 0$, *there exists* $\delta > 0$, *such that*[2]

$$0 < |x - a| < \delta \implies |f(x) - l| < \epsilon. \tag{5.5}$$

We verify this criterion for the limit claimed in (5.2) Suppose $\epsilon > 0$ is given. Note that for $x \neq 3$

$$|y - 7| = |(2x + 1) - 7| = 2|x - 3|.$$

So choosing $\delta = \epsilon/2$, we have

$$0 < |x - 3| < \delta \implies |y - 7| < \epsilon$$

as required.

Non-Existence of the Limit

If the graph of a function has a *jump* at a certain point (rather than a hole), then the limit of the function will fail to exist at that point. Consider, for example, the function whose graph is shown in Figure 5.5.

As x approaches 4 from the left (meaning we are on the lower branch of the graph) the values of y approach 9, whereas, as x approaches 4 from the right, y approaches 12. We express this by saying the *left* and *right* limits at 4 exist, and denote them by

$$\lim_{x \to 4^-} y = 9, \qquad \lim_{x \to 4^+} y = 12.$$

[2]The restriction $0 < |x - a|$ in 5.5 is significant. It allows for the limit to exist at $x = a$ even when $f(a)$ is undefined.

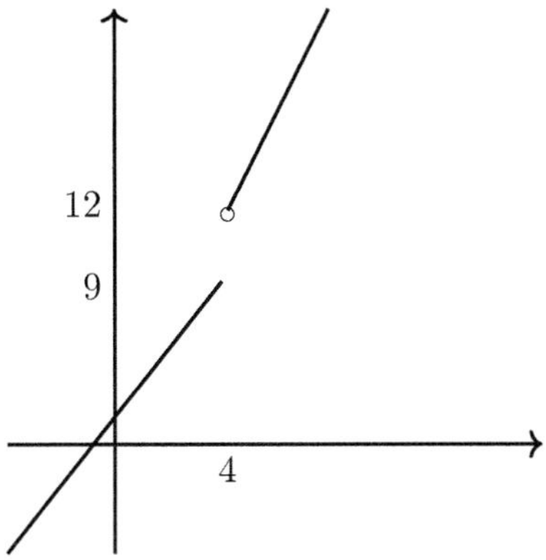

FIGURE 5.5 A graph with a jump discontinuity

Since the two one-sided limits differ, $\lim_{x\to 4} y$ does not exist, as will now demonstrate. The proof is a short contradiction argument.

Suppose that, on the contrary,

$$\lim_{y\to 4} = l. \tag{5.6}$$

Apply the definition of limit with $\epsilon = 1$. Then there exists $\delta > 0$ such that

$$0 < |x - 4| < \delta \implies |y - l| < 1, \text{ i.e., } l - 1 < y < l + 1. \tag{5.7}$$

Choose any two points $x_1 \in (-\delta, 4)$ and $x_2 \in (4, \delta)$. By (5.7), both $y(x_1)$ and $y(x_2)$ lie in the interval $(l - 1, l + 1)$ and therefore differ by less than 2. But this is impossible: since x_1 and x_2 fall on opposite sides of 4, the gap in y at these two points must be at least 3.

A further definition is useful.

Definition. Suppose both $f(a)$ and $\lim_{x\to a} f(x)$ exist, and

$$\lim_{x\to a} f(x) = f(a).$$

Then we say f is *continuous at* a. If f is continuous at all points in its domain, then f is said to be a *continuous* function.

Continuity has an obvious graphical meaning: it is equivalent to the graph of the function being a continuous (i.e., unbroken) curve at the point in question.

Exercise 5.2.[⋆] *Determine the values of c so that the piecewise continuous function*

$$f(x) = \begin{cases} x^2 + 2c, & \text{if } x < 3 \\ 3x + c^2, & \text{if } x \geq 3 \end{cases}$$

is continuous at $x = 3$.

Theorem 5.1. *Polynomial functions* $f(x) = a_o + a_1x + a_2x^2 + \cdots + a_nx^n$ *are continuous. Rational functions (functions of the form* $p(x)/q(x)$, *where* p *and* q *are polynomials), are continuous except at values of* x *where their denominators are zero.*

For example, the function

$$f(x) = \frac{x^2 + 1}{x^2 - 4}$$

is continuous for all x except ± 2.

We close this section by stating one of the most basic properties of continuous functions, known as the Intermediate Value Theorem (IVT).

Theorem 5.2. *Suppose* f *is a continuous function defined on an interval* $[a, b]$ *and let* d *be a value lying between* $f(a)$ *and* $f(b)$. *Then there exists* $c \in [a, b]$ *such that* $f(c) = d$.

IVT, which seems obvious graphically, is actually a consequence of a deep and subtle assumption about the real number system known as *completeness.* As an amusing application of IVT, we can prove that at any given time, there is at least one place in the world where the temperature is exactly $0°C$. Draw any curve on the globe starting from the North Pole (N) and ending on the Equator (E) and consider the temperature T along this curve Then T varies continuously as we move along this curve. Furthermore, $T(N) < 0$ and $T(E) > 0$. So, by IVT, there exists a point p on the curve between N and E where $T(p) = 0$.

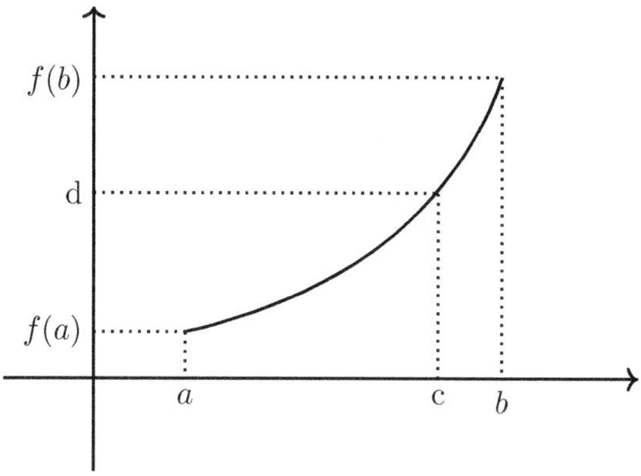

FIGURE 5.6 The Intermediate Value Theorem

5.3 THE DERIVATIVE

Consider the following example. Given the function $y = x^2$, determine the slope of the tangent line at the point $x = 3$. There is an obvious difficulty: to compute the slope of a line, we need two points on the line and here we have only the single point $(3, 9)$ (call it P). We create a second point Q on the curve $(3 + h, (3 + h)^2)$, as indicated in figure 5.7.

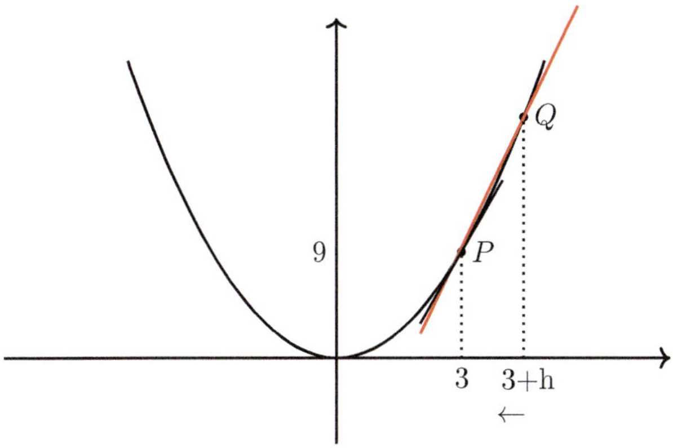

FIGURE 5.7

The slope of the secant line PQ is

$$\frac{(3+h)^2 - 9}{3+h-3} = \frac{6h + 6h^2}{h} = 6 + h.$$

We now take a limit as h approaches 0. The point Q approaches P and the secant line PQ becomes the tangent line at P. Thus the slope of the tangent line is given by

$$\lim_{h \to 0} 6 + h = 6.$$

More generally, this process can be used to calculate the slope of the tangent line to an arbitrary curve $y = f(x)$. The slope of the secant line is the expression (called the *difference quotient*)

$$\frac{f(x+h) - f(x)}{h}.$$

As before, the slope of the tangent line is found by taking the limit as $h \to 0$.

This motivates the following definition. We say the function f is *differentiable* at x if the limit

$$\lim_{h \to 0} \frac{f(x+h) - f(x)}{h}$$

exists. The limit function is called the *derivative* of f and denoted by $f'(x)$ or df/dx.

The derivative is one of the two primary operations in calculus. As we have seen, it gives the slope of the tangent line to the graph of a function and thereby measures the rate of change of the function. If the function represents the position of a moving object then the derivative gives the velocity of the object.

Example

A diver jumps from a platform that is 32 feet above the water with an initial velocity of 32 feet per second. The height of the diver at time t is given by the function

$$s(t) = -16t^2 + 16t + 32.$$

After how many seconds will the diver hit the water? Find the diver's velocity at impact. What is the maximum height of the diver?

We start by calculating and simplifying the difference quotient.

$$\begin{aligned} &\frac{s(t+h) - s(t)}{h} \\ &= \frac{-16(t+h)^2 + 16(t+h) + 32 - (-16t^2 + 16t + 32)}{h} \\ &= \frac{-32ht - 16h^2 + 16h}{h} \\ &= \frac{h(-32t - 16h + 16)}{h} \\ &= -32t + 16h + 16. \end{aligned}$$

Hence the velocity $v(t)$ of the diver at time t is

$$v(t) = \lim_{h \to 0} -32t + 16h + 16 = -32t + 16 \text{ ft/sec.}$$

The diver hits the water when

$$-16t^2 + 16t + 32 = 16(2 - t)(t + 1) = 0,$$

i.e., after time $t = 2$ seconds (we disregard the negative solution as unfeasible). The velocity on impact is $v(2) = -48$ ft/sec. Finally, the maximum height is attained when $v(t) = 0$, i.e., $t = 1/2$, and $s(1/2) = 36$ ft.

Differentiability Versus Continuity

The definition of derivative calls for the limit of the difference quotient as $h \to 0$. If the limit does not exist at a point a, then the function is said to be *non-differentiable* at a. In general, differentiability at a point implies continuity

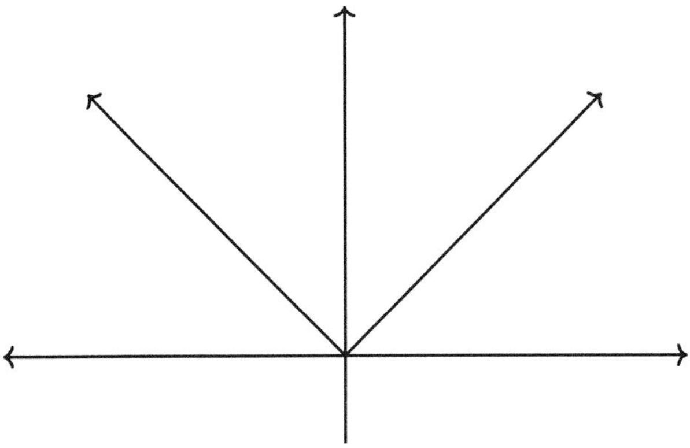

FIGURE 5.8 Graph of $y = |x|$

there, but not conversely. Consider, for example, the absolute value function $y = |x|$.

The difference quotient (DQ) at $x = 0$ is

$$\frac{|0 + h| - |0|}{h} = \frac{|h|}{h}.$$

Recall that absolute value is actually a piecewise function, defined by

$$|x| = \begin{cases} -x, & \text{if } x < 0, \\ x, & \text{if } x \geq 0. \end{cases}$$

Thus, for $h < 0$, DQ $= -h/h = -1$, while for $h > 0$, DQ $= h/h = 1$. Consequently $\lim_{h \to 0^-} -1 = -1$ and $\lim_{h \to 0^+} 1 = 1$. Because the two one-sided limits are different, the limit of the difference quotient as $h \to 0$ (i.e., the derivative of $|x|$ at $x = 0$) does not exist. In short, the function $|x|$ is continuous, but non-differentiable at 0.

Note the sharp corner in the graph $y = |x|$ (figure 5.8) at $x = 0$. For this reason, it is not possible to draw a (unique) tangent line to the graph at $x = 0$. In general, when a continuous function fails to be differentiable at a point, the graph will have either a vertical tangent line or a corner (or a cusp) at the point.

Differentiation Rules

Before getting to this let's work another example. The problem is to calculate the derivative of the function $y = \sqrt{x}$, $x > 0$. We start with the difference quotient, which we simplify by rationalizing the numerator as before, to obtain

$$\begin{aligned}\frac{\sqrt{x+h}-\sqrt{x}}{h} &= \frac{\sqrt{x+h}-\sqrt{x}}{h} \times \frac{\sqrt{x+h}+\sqrt{x}}{(\sqrt{x+h}+\sqrt{x})} \\ &= \frac{(\sqrt{x+h})^2-(\sqrt{x})^2}{h(\sqrt{x+h}+\sqrt{x})} = \frac{h}{h(\sqrt{x+h}+\sqrt{x})} \\ &= \frac{1}{\sqrt{x+h}+\sqrt{x}}.\end{aligned}$$

Thus the derivative of $\sqrt{x}$ is given by

$$\lim_{h\to 0}\frac{1}{\sqrt{x+h}+\sqrt{x}} = \frac{1}{2\sqrt{x}}. \tag{5.8}$$

Calculating the derivatives by the basic method for more complicated functions would be an onerous task. Fortunately there are formulas which allow us to do this much more easily. These go by the name of *differentiation rules.* The main ones are as follows.

Theorem 5.3.

(*i*) *(Power rule) For every rational power, the function x^r has derivative rx^{r-1}.*
(*ii*) *(Addition rule)* $(f+g)' = f' + g'$.
(*iii*) *(Product rule)* $(f\cdot g)' = f'g + fg'$.
(*iv*) *(Quotient rule)* $\left(\frac{f}{g}\right)' = \frac{f'g - fg'}{g^2}$.
(*v*) *(Chain rule)* $(g\circ f)'(x) = g'(f(x)f'(x)$.

As an example of Theorem 5.3(i) we obtain a much easier derivation of (5.8)

$$\frac{d}{dx}\sqrt{x} = \frac{d}{dx}x^{1/2} = \frac{1}{2}x^{-1/2} = \frac{1}{2\sqrt{x}}.$$

Theorem 5.3 provides for the differentiation of rational functions. Consider, e.g.,

$$y = \frac{3x^2-1}{x^2+4}.$$

Using Theorem 5.3(iv), we obtain

$$\begin{aligned}y' &= \frac{(3x^2-1)'(x^2+4)-(3x^2-1)(x^2+4)'}{(x^2+4)^2} \\ &= \frac{6x(x^2+4)-(3x^2-1)2x}{(x^2+4)^2} \\ &= \frac{14x}{(x^2+4)^2}.\end{aligned}$$

Exercise 5.3.* *Use the Quotient Rule (Theorem 5.3(iv)) to find the equation of the tangent line to the graph*

$$y = \frac{(1 - 2x + 3x^2)(2 + x - x^3)}{1 + 4x + 3x^2}$$

at $x = 1$.

The Chain Rule (Theorem 5.3(v)) has many important corollaries. One of them is the following.

Theorem 5.4. *(Inverse Function Theorem) Suppose* f *and* g *are inverse functions,* f *is differentiable at* $g(y)$ *and* $f'(g(y)) \neq 0$. *Then* g *is differentiable at* x *and*

$$g'(y) = \frac{1}{f'(g(y))}. \tag{5.9}$$

Exercise 5.4.* *Use the Inverse Function Theorem (IFT) to calculate* $(f^{-1})'(21)$, *where*

$$f(x) = x^3 + 4x + 5.$$

Exercise 5.5. *Use the Inverse Function Theorem to calculate* $(f^{-1})'(7)$, *where*

$$f(x) = \frac{3x + 5}{4x - 5}.$$

Do this in two ways, by solving for the inverse function, and by using IFT.

The following formulas give the derivatives of the basic trigonometric functions.

Theorem 5.5.

$$\begin{aligned}
&\textit{(i)} && \sin'(x) = \cos x, \\
&\textit{(ii)} && \cos'(x) = -\sin x, \\
&\textit{(iii)} && \tan'(x) = \sec^2 x.
\end{aligned}$$

A Strange and Striking Result

Theorem 5.6. *Suppose* $f = F'$ *for some differentiable function* F. *Then* f *has the Intermediate Value Property (i.e., assumes every value between any two values that it takes).*

A *discontinuous* function need not have the Intermediate Value Property (IVP). An obvious example is a function with a "jump discontinuity" (see figure 5.5 where y never takes the value 10.) Theorem 5.6 is fascinating in light of the fact that the derivative of a function need not be continuous! This behavior is illustrated by the example

$$F(x) = \begin{cases} x^2 \sin 1/x, & \text{if } x \neq 0, \\ 0, & \text{if } x = 0. \end{cases}$$

Then

$$f(x) = F'(x) = \begin{cases} 2x \sin 1/x - \cos 1/x, & \text{if } x \neq 0, \\ \lim_{h \to 0} h \sin 1/h = 0, & \text{if } x = 0. \end{cases}$$

The function f performs infinitely many oscillations as $x \to 0$. Graph it with a calculator or a computer to see. Thus f is wildly discontinuous at 0, yet has the IVP (and in spades).

5.4 THE INTEGRAL

A rigorous basis for integration was established in the nineteenth century by several mathematicians, most notably Darboux and Riemann. (Later Lebesgue created a more general integration theory.) As remarked earlier, the integral arises when computing the area under a given curve. If the curve is prescribed using analytic geometry, this is no big deal—at least if only an approximate answer is required; then the integral can approximated by (a finite number of) rectangles. But the determination of a *precise* value for the area requires the development of theory involving limits.

Whereas the limits discussed previously were those where the independent variable x approaches a finite point, the limits involved in integration are those at infinity. These are construed as follows. To say that the sequence a_n has limit L as n tends to infinity (denoted $\lim_{x\to\infty} a_n = L$), means that a_n approaches arbitrarily close to L as the index n becomes large. As an example, consider

$$a_n = \left(1 + \frac{1}{n}\right)^n$$

(Formulas of this type occur in calculating compound interest.) Then $a_{10} = 2.59374...$, $a_{100} = 2.70481...$, $a_{1000} = 2.71692...$, $a_{10^6} = 2.71828...$ We observe that as n increases indefinitely, a_n gets ever closer to a finite quantity, 2.71828182845... This is the number e introduced in the next section. We say

$$\lim_{n\to\infty} \left(1 + \frac{1}{n}\right)^n = e.$$

Upper and Lower Sums

Suppose it is required to compute the area A lying beneath the curve $y = f(x), a \leq x \leq b$.

The Darboux approach to the problem is as follows. For an arbitrary integer n, let Δx denote the quantity $(b - a)/n$. Define points $x_0 = a$, $x_1 =$

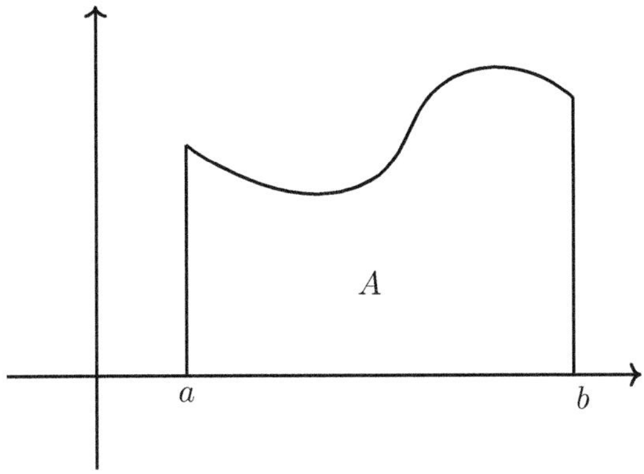

FIGURE 5.9 Region under a curve

$a + \Delta x$, $x_2 = a + 2\Delta x$, $x_3 = a + 3\Delta x, \ldots, x_n = a + n\Delta x = b$. Thus $x_0, x_1, x_2, \ldots, x_n$ are $n+1$ equally spaced points starting at a and ending at b. (This arrangement is called a *regular partition* of the interval $[a, b]$.) On each subinterval $[x_{k-1}, x_k]$ draw a *lower* rectangle choosing for the height of the rectangle, the *minimum* value m_k of the function f on the interval $[x_{k-1}, x_k]$, and an *upper* rectangle choosing for the height, the *maximum*[3] value of f on $[x_{k-1}, x_k]$ (see figure 5.10).

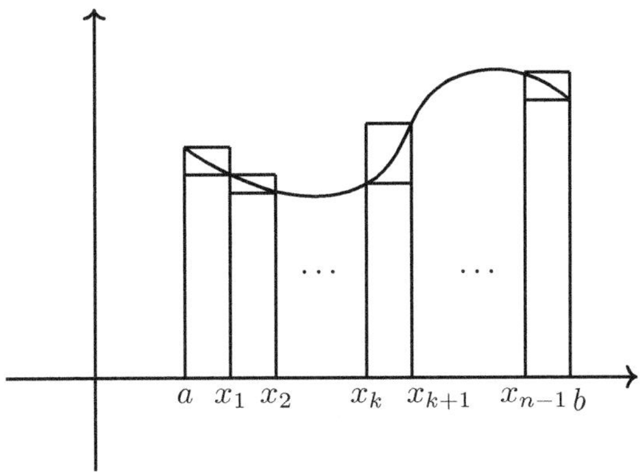

FIGURE 5.10 Upper and lower rectangles

Denote the sums of the areas of the upper and lower rectangles by U_n and

[3]In order to simplify the exposition, we are assuming f assumes maximum and minimum values on the intervals, but there is a more general formulation.

L_n respectively ("upper" and "lower" sums). In "sigma" notation

$$L_n = \sum_{k=1}^{n} m_k \Delta x,$$
$$U_n = \sum_{k=1}^{n} M_k \Delta x.$$

Since the totality of the upper rectangles contains the region under the curve, while the reverse is true for the lower rectangles, we have, for each n, the inequalities

$$L_n \leq A \leq Un. \tag{5.10}$$

As n increases, the lower sums become larger and the upper sums smaller. Suppose it happens that as n tends to infinity, L_n and U_n approach a common limit (call it L). Then it follows by (5.10) that $A = L$. In this situation the sums provide approximations to the area A and if we can compute the limit of the sums as n tends to infinity, the precise value of A.

Definition. Suppose that, in the above notation

$$\lim_{n\to\infty} L_n = \lim_{n\to\infty} U_n = L. \tag{5.11}$$

Then the function $f(x), a \leq x \leq b$ is said to be *integrable* and we define the (definite) integral

$$\int_a^b f(x)dx = L.$$

As remarked earlier, this notation and conception[4] of the integral originated with Leibniz. The integral sign $\int$ is an elongated "S," introduced by Leibniz to signify sum.

We compute the example

$$\int_0^1 x^2 dx.$$

Partition the interval $[0, 1]$ into points

$$x_0 = 0,\ x_1 = \frac{1}{n},\ x_2 = \frac{2}{n}, \ldots, x_{n-1} = \frac{n-1}{n},\ x_n = 1.$$

Since the function is increasing, the maximum (resp. minimum) on each interval is taken at the right (resp. left) hand endpoint. Thus the upper and lower

[4]By contrast Newton's *fluxions* signified the alternative interpretation of the integral as an antiderivative. The two different notions of integral are reconciled by the Fundamental Theorem.

sums are

$$U_n = \left[\left(\frac{1}{n}\right)^2 + \left(\frac{2}{n}\right)^2 + \left(\frac{3}{n}\right)^2 + \cdots + \left(\frac{n}{n}\right)^2\right]\frac{1}{n} = \frac{1}{n^3}\sum_{k=1}^{n} k^2,$$

$$L_n = \left[\left(\frac{0}{n}\right)^2 + \left(\frac{1}{n}\right)^2 + \left(\frac{2}{n}\right)^2 + \cdots + \left(\frac{n-1}{n}\right)^2\right] = \frac{1}{n^3}\sum_{k=1}^{n-1} k^2.$$

Using the summation formula

$$\sum_{k=1}^{n} \frac{1}{k^2} = \frac{n(n+1)(2n+1)}{6}$$

we obtain

$$U_n = \frac{n(n+1)(2n+1)}{6n^3} = \frac{1}{3} + \frac{1}{2n} + \frac{1}{6n^2},$$

$$L_n = \frac{(n-1)n(2[n-1]+1)}{6n^3} = \frac{1}{3} - \frac{1}{2n} + \frac{1}{6n^2}.$$

Note that terms with n in the denominators have limit 0. Thus $\lim_{n\to\infty} U_n = \lim_{n\to\infty} L_n = 1/3$. We conclude that the function is integrable and

$$\int_0^1 x^2 = \frac{1}{3}.$$

Exercise 5.6. *Show that the function*

$$y = 2x + 1, \; 1 \le x \le 3$$

is integrable and compute

$$\int_1^3 2x + 1 \, dx$$

as a limit of either upper or lower sums. Verify this result using the formula for the area of a Trapezoid.

An Example of a Non-Integrable Function

The standard example is the so called *Dirichlet function*. This is the function f defined on the interval $[0,1]$ by $f(x) = 1$ if x is a rational number and 0 if x is an irrational number (see figure 5.11).

It happens that every interval (a,b) contains both rational points and irrational points (this property is referred to as *density* of both the rationals and irrationals). Thus the maximum M_k of f on each of the intervals $[(k-1)/n, k/n]$, is 1 and the minimum m_k is 0. This implies every upper sum

$$U_n = \frac{1}{n}\sum_{k=1}^{n} 1 = 1$$

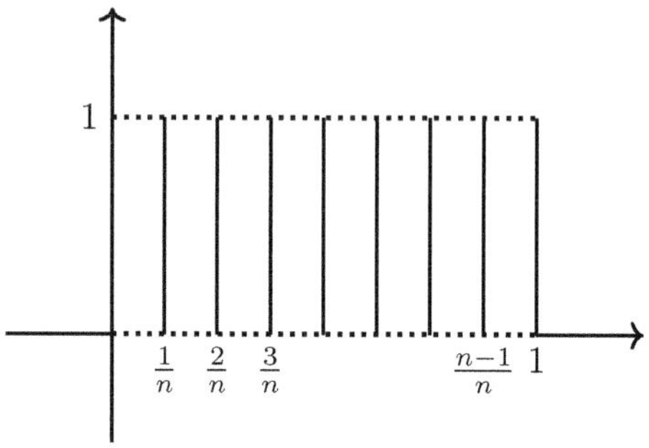

FIGURE 5.11 The Dirichlet function

while every lower sum

$$L_n = \frac{1}{n}\sum_{k=1}^{n} 0 = 0.$$

Hence $\lim_{n\to\infty} U_n = 1$ and $\lim_{n\to\infty} L_n = 0$ and so the function is non-integrable.

The non-integrability of the Dirichlet function is clearly related to its highly irregular behavior, discontinuous on every interval. The following result puts this example in context.

Theorem 5.7. *Suppose f is a continuous function defined on an interval $[a, b]$, then f is integrable. More generally, the same is true for every piecewise continuous function.*[5]

5.5 THE FUNDAMENTAL THEOREM OF CALCULUS

There are actually two such theorems, part I and part II. The first of these is as follows.

The First Fundamental Theorem

Theorem 5.8. *FTC(I) Suppose $f : [a, b] \mapsto \mathbf{R}$ is an integrable function and F is an antiderivative of f (i.e., a function F defined on $[a, b]$ such that $F' = f$). Then*

$$\int_a^b f(x)dx = F(b) - F(a). \tag{5.12}$$

[5] A function whose graph consists of a finite number of unbroken pieces.

FTC(I) is a most powerful tool for computing integrals. By way of illustration we use it to evaluate the integral

$$\int_0^1 x^2 dx,$$

which we calculated earlier. First note that the function in question is integrable by Theorem 5.7.

Antiderivatives for powers x are easily found by reversing the rule for differentiation. Thus for $p \neq -1$, we see that an antiderivative[6] of x^p is

$$\frac{x^{p+1}}{p+1}.$$

In particular, $F(x) = x^3/3$ is an antiderivative of x^2. By FTC(I) we have

$$\int_0^1 x^2 dx = F(1) - F(0) = \frac{1^3}{3} - \frac{0^3}{3} = \frac{1}{3}.$$

Exercise 5.7. *Use FTC(I) to calculate the area enclosed between the graph $y = 4 - x^2$ and the x-axis and show that your answer agrees with Archimedes' area for a parabolic segment given in Section 1.4.*

The Second Fundamental Theorem of Calculus

The second part of the fundamental theorem is as follows.

Theorem 5.9. *FTC(II). Suppose the function $f : [a, b] \mapsto \mathbf{R}$ is continuous. Consider the function*

$$F(x) = \int_0^x f(t)dt, \quad a < x < b.$$

Then F is an antiderivative of f, i.e.,

$$\frac{d}{dx}\int_0^x f(t)dt = f(x).$$

This theorem establishes integration and differentiation as *inverse* operations (i.e., performing the first operation on a continuous function and then the other takes you back to the function).

It should be noted that the assumption of continuity of f in the theorem is essential. To see this consider the following example.

$$s(t) = \begin{cases} 0, & \text{if } 0 \leq t \leq 1, t \neq 1/2, \\ 1, & \text{if } t = 1/2. \end{cases}$$

[6]Antiderivatives are clearly not unique, e.g., $\frac{x^3}{3} + 1$ is also an antiderivative of x^2. However, it can be shown that any two antiderivatives differ only by a constant. The constant is redundant in FTC(I) since it cancels out in the subsequent subtraction.

For x in the range $0 \leq x < 1/2$, we have $L_n = U_n = 0$ for all n, so

$$S(x) = \int_0^x s(t)dt$$

exists and equals 0. For $1/2 \leq x \leq 1, L_n = 0$ and $U_n = 1/n$ (see figure 5.12, so again the integral exists and equals 0. In short, $S \equiv 0$ (is zero everywhere on $[0, 1]$). In particular $S'(1/2) = 0 \neq s(1/2)$.

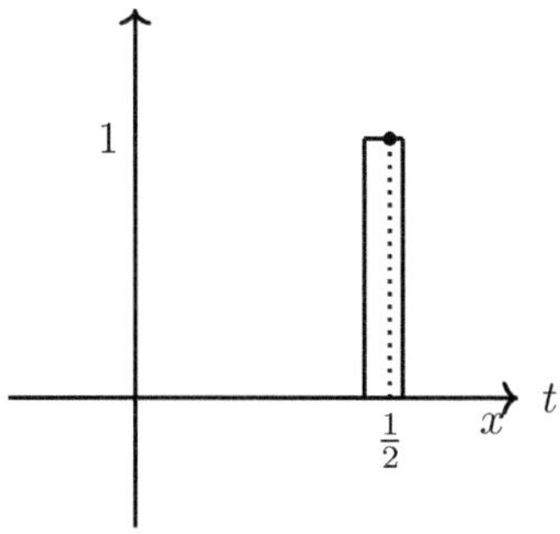

FIGURE 5.12 Single point function

Combining FTC(II) with the Chain Rule, we obtain the following result.

Theorem 5.10. *Suppose f is continuous and g is differentiable. Define*

$$y = \int_a^{g(x)} f(t)dt.$$

Then y is differentiable and

$$y' = f(g(x))g'(x).$$

Proof. Simply apply the Chain Rule to the composition $F \circ g$ and use FTC(II), where

$$F(x) = \int_a^x f(t)dt.$$

□

Exercise 5.8. *Consider the function*

$$y = \int_{\cos x}^{\sin x} \sqrt{1 - t^2}dt = \int_0^{\sin x} \sqrt{1 - t^2}dt - \int_0^{\cos x} \sqrt{1 - t^2}dt.$$

Use Theorem 5.10 to show that $y' \equiv 1$ and hence y has the form $x + c$. By substituting $x = 0$ and then $x = \pi/2$, deduce that

$$\int_0^1 \sqrt{1 - t^2}dt = \frac{\pi}{4},$$

i.e., the area of a circle of radius 1 is π.

The Natural Logarithm and Exponential Functions

Up to this point there has been little mention in the book of exponential or logarithm functions. Our reason for leaving it to now is that these functions are most naturally introduced using integration.

The natural logarithm of x (written $\log x$) is defined by

$$\log x = \int_1^x \frac{1}{t}\,dt, \quad x > 0.$$

Certain properties of log are immediately apparent from the definition:
(i) $\log 1 = 0$.
(ii) log is a strictly increasing function, i.e.,

$$x_2 > x_1 \implies \log x_2 > \log x_1. \tag{5.13}$$

(This is clear from the interpretation of the integral as area under the graph.)
(iii) From FTC(II), the function $\log x$ is differentiable, with derivative

$$\log' x = \frac{1}{x}. \tag{5.14}$$

Wie can establish the characteristic log properties: for all $x, y > 0$ and rational powers p:

$$\begin{aligned} \log x + \log y &= \log xy, \\ p \log x &= \log x^p. \end{aligned} \tag{5.15}$$

To show the first of these (the second follows in similar fashion), fix $y > 0$ and consider the function $f(x) \equiv \log xy$. By Theorem 5.13 we have

$$f'(x) = \frac{1}{xy} \cdot (xy)' = \frac{y}{xy} = \frac{1}{x} = \log'(x).$$

It follows that there exists a constant c such that

$$\log xy = \log x + c.$$

Evaluating at $x = 1$ gives $c = \log y$.

The strictly increasing behavior of $\log x$ noted above implies that there exists an inverse function. We denote the inverse function (temporarily) by $\exp(x)$. This function has derivative

$$\exp'(x) = \exp(x). \tag{5.16}$$

A function that is immune to differentiation!

$$\begin{aligned} \exp(x)\exp(y) &= \exp(x + y), \\ \exp(px) &= [\exp(x)]^p. \end{aligned} \tag{5.17}$$

Exercise 5.9* *Prove (5.16) using (5.14) and the Inverse Function Theorem.*

Exercise 5.10. *Prove (5.17) using (5.15).*

Denote $\exp(1)$ by e. From the first part of (5.17) we have $\exp(2) = \exp(1+1) = \exp(1) \cdot \exp(1) = e^2$. Similarly, for any positive integer n, $\exp(n) = e^n$. Using the second part of (5.17) we have $\left[\exp(n/m)\right]^m = \exp(n) = e^n$. Thus

$$\exp(n/m) = e^{n/m}.$$

In words, $\exp(x)$ is a continuous extension of the function e^x from the rational numbers to *all real numbers*. For this reason, it is customary to write e^x in place of $\exp(x)$ and we will do so from now on.

So, what is this number e? From the designation $e = \exp(1)$ and $\log x$ and $\exp(x)$ as inverse functions, we have $\log e = 1$. That is, e the number such that

$$\int_1^e \frac{1}{t}\,dt = 1.$$

We can get a very rough estimate of e from figure 5.13. The figure shows that $\log 2 < 1$, while $\log 4 > 1$. Thus $2 < e < 4$. More rectangles give better estimates. A calculator gives $e = 2.71828182845....$

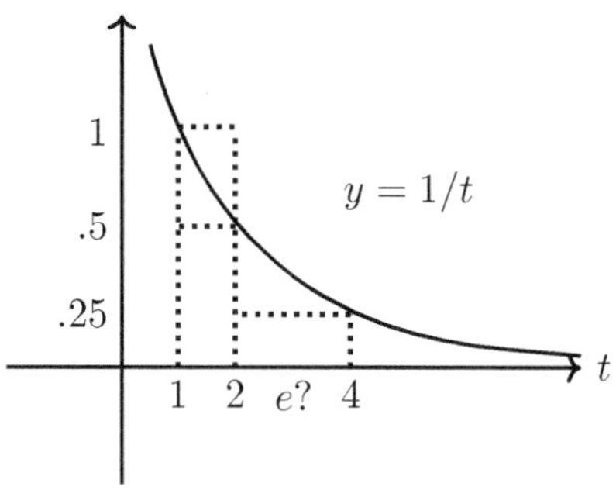

FIGURE 5.13 Crude estimation of e

5.6 TAYLOR SERIES

Functions such as

$$f(x) = \sqrt[3]{\frac{x^3 - 8}{3x^2 + 7}},$$

which can be computed from the arithmetic operations and root extractions in a finite number of steps are said to be *algebraic*. Functions that cannot be expressed in this way are termed *transcendental*. The class of transcendental functions includes the trigonometric functions, $\log x$ and e^x. Transcendental functions, by definition, defy computation in a finite number of steps. The best that one can hope for is to come up with an approximation. The most easily computed expressions are polynomials.

The equation of the tangent line to the curve $y = f(x)$ at $x = c$,

$$y = f(c) + f'(c)(x - c) \tag{5.18}$$

provides the best linear approximation to the curve in a neighborhood of c (see figure 5.14).

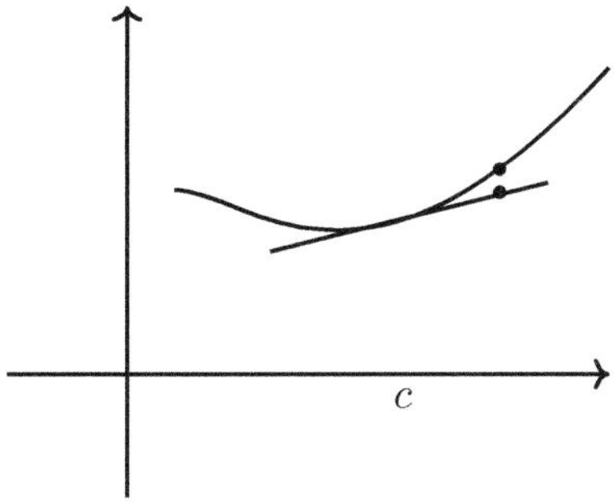

FIGURE 5.14 Approximation by the tangent line

To take an example, suppose it is required to find an approximation for $\sin 39°$. Choose for c a point close to $39°$ at which we can compute the tangent line (i.e., where we know the sine and cosine). A good choice would be $45°$ (in radians $\pi/4$). The tangent line to $y = \sin x$ at $x = \pi/4$ is

$$y = \frac{\sqrt{2}}{2} + \frac{\sqrt{2}}{2}\left(x - \frac{\pi}{4}\right).$$

Substituting $x = .68...$ (the radian measure of $39°$) gives $y = .6325...$ The calculator value for $\sin 39°$. is $.6287...$

The reason why, of all straight lines passing through the point $(c, f(c))$, the tangent line at c stays closest to the curve in a neighborhood of c is that its slope $f'(c)$ agrees with that of the curve at $x = c$. However, we do not need to restrict ourselves to a linear approximation. Suppose that, instead of the tangent line at c, we approximate f by a quadratic, a cubic, or any degree-n polynomial P_n. Provided f is n-times[7] differentiable, we can choose P_n so that the polynomial, together with its higher-order derivatives up to nth order, agree with those of f at c, i.e.,

$$P_n(c) = f(c),\ , P_n'(c) = f'(c),\ P_n''(c) = f''(c), \ldots, P_n^{(n)}(c) = f^n(c). \tag{5.19}$$

[7]The higher order derivatives of a function f, denoted ether by several primes or by the superscript are obtained by successive differentiation. For example, for $f(x) = x^3$, we have $f'(x) = 3x^2$, $f''(x) = 6x$, $f'''(x) = 6$, $f^{(k)}(x) = 0$ for $k \geq 4$.

Theorem 5.11. *Suppose the function is n-times differentiable at c. Then there exists a unique degree-n polynomial P_n satisfying (5.19). It is the polynomial*[8]

$$P_n(x) = \sum_{k=1}^{n} \frac{f^{(k)}(c)}{n!}(x-c)^k. \tag{5.20}$$

Exercise 5.11. *Prove the polynomial (5.20) satisfies (5.19).*

The polynomial (5.20) is known as the degree-n *Taylor Polynomial* of f centered at c, named for the English mathematician Brook Taylor. In the special case $c = 0$, it is referred to as the *Maclaurin polynomial.* The tangent line is the first-degree case $n = 1$.

Maclaurin Polynomials for the Basic Transcendental Functions

We derive the Maclaurin polynomials for the exponential, sine, cosine, and natural logarithm functions.

Since the function $f(x) = e^x$ has derivative e^x, it follows that the same holds for the derivatives of all orders. Hence $f^{(k)}(0) = e^0 = 1$ for all $k \geq 0$, and we have

$$P_n(x) = 1 + x + \frac{x^2}{2!} + \frac{x^3}{3!} + \cdots + \frac{x^n}{n!}. \tag{5.21}$$

The derivatives of the sine function exhibit a cyclical pattern. For $f(x) = \sin x$

$$\begin{aligned} f'(x) &= \cos x \\ f''(x) &= -\sin x \\ f^{(3)}(x) &= -\cos x \\ f^{(4)}(x) &= \sin x \end{aligned}$$

and this pattern repeats itself for the next 4 derivatives, and then the next 4, etc. Since $\sin 0 = 0$ and $\cos 0 = 1$, we have for the sine function and odd n,

$$P_n(x) = x - \frac{x^3}{3!} + \frac{x^5}{5!} - \frac{x^7}{7!} + \ldots (\pm)\frac{x^n}{n!}. \tag{5.22}$$

When n is even, since $f^{(n)}(0) = 0$, $P_n = P_{n-1}$.

A similar argument applies to cosine. In this case, the Maclaurin series is, for n even,

$$P_n(x) = 1 - \frac{x^2}{2!} + \frac{x^4}{4!} - \frac{x^6}{6!} + \cdots (\pm)\frac{x^n}{n!}. \tag{5.23}$$

(For n odd, $P_n = P_{n-1}$.)

[8] In (5.15) we are using the conventions $f^{(0)} = f$ and $0! = 1$.

Since $\log 0$ is undefined, we compute the Maclaurin polynomial for the function $f(x) = \log(1+x)$. The first few derivatives in this case are as follows:

$$\begin{aligned}
f'(x) &= \frac{1}{1+x} \\
f''(x) &= -\frac{1}{(1+x)^2} \\
f^{(3)}(x) &= \frac{2}{(1+x)^3} \\
f^{(4)}(x) &= -\frac{6}{(1+x)^4} = -\frac{3!}{(1+x)^4} \\
f^{(5)}(x) &= \frac{24}{(1+x)^5} = \frac{4!}{(1+x)^5}.
\end{aligned}$$

It is clear that the pattern emerging here is

$$f^{(k)}(x) = (-1)^{k-1} \frac{(k-1)!}{(1+x)^k}. \tag{5.24}$$

Since $f(0) = 1$, we have in this case

$$\begin{aligned}
\log(1+x) &= x - \frac{x^2}{2} + \frac{2x^3}{3!} - \frac{3!x^3}{4!} + \cdots (\pm)\frac{(k-1)!x^k}{k!} \\
&= x - \frac{x^2}{2} + \frac{x^3}{3} - \frac{x^4}{4} + \cdots (\pm)\frac{x^k}{k}.
\end{aligned} \tag{5.25}$$

Taylor's Theorem

Taylor polynomials provide a natural way to approximate transcendental functions. But how accurate is the approximation? An answer to this question is provided by the following result, known as Taylor's theorem (though this particular version of the theorem is due to Lagrange).

Theorem 5.12. *Suppose f is $k+1$ times differentiable on the interval (a, x) and $f^{(n+1)}$ is continuous on $[a, x]$. Then*

$$f(x) = P_n(x) + R_k(x),$$

where P_n is the Taylor polynomial (5.20) and $R_n(x)$(called the remainder, or error term) is given by

$$R_n(x) = \frac{f^{(n+1)}(\xi)}{(n+1)!}(x-c)^{n+1},$$

where ξ is a number between x and c.

As an example in the use of Taylor's theorem, suppose it is required to compute an approximation to $\log 1.2$ accurate to within 0.000001 (10^{-6}). According to the theorem applied to $\log(1+x)$ with $c = 0$, and (5.24) the remainder term here has the form

$$R_n = \frac{(-1)^n(0.2)^{n+1}}{(n+1)(1+\xi)^{n+1}},$$

where $0 < \xi < 0.2$. Since

$$|R_n| \leq \frac{(0.2)^{n+1}}{n+1},$$

it suffices to choose n such that the latter quantity is less than 10^{-6}, inspection shows that $n = 10$ will do, then compute the Maclaurin polynomial

$$P_{10}(0.2) = 1 - 0.2 + \frac{(0.2)^2}{2} - \frac{(0.2)^3}{3} + \frac{(0.2)^4}{4} - \frac{(0.2)^5}{5} + \frac{(0.2)^6}{6} - \frac{(0.2)^7}{7} + \frac{(0.2)8}{8} - \frac{(0.2)^9}{9} + \frac{(0.2)^{10}}{10}.$$

Exercise 5.12. *Use Taylor's theorem to compute an approximation for* $\sqrt[3]{8.2}$ *accurate to within* 0.000001.

In general, the error term $R_n(x)$ gets smaller as n gets larger, thus $P_n(x)$ gives a better approximation to f (at least locally, close to $x = c$). What can we expect if we let $n \to \infty$? Well, for one thing, the Taylor polynomial gives rise to an infinite sum of terms known as a *Taylor series.* To make sense of the notion of "infinite sum" we must once again evoke the limit.

Suppose $a_0, a_1, a_2, \ldots$ is sequence of numbers. We define the nth *partial sums* of the sequence

$$S_n = \sum_{k=1}^{n} a_n.$$

If $\lim_{n\to\infty} S_n = S$ exists then we say the series

$$\sum_{k=1}^{\infty} a_n$$

is *convergent*, with sum S. Otherwise, the series is said to be *divergent.*

For example, there is the well-known formula for the sum of the geometric series

$$\sum_{k=0}^{n-1} r^k = \frac{1-r^n}{1-r}, \qquad r \neq 1.$$

It is straightforward to show that $\lim_{n\to\infty} r^n = 0$, if $-1 < r < 1$. Hence, for r in this range the (infinite) geometric series is convergent and

$$\sum_{k=0}^{\infty} r^k = \frac{1}{1-r}. \tag{5.26}$$

In general, if it can be shown that $\lim_{n\to\infty} R_n(x) = 0$, then it follows that

$$f(x) = \sum_{k=1}^{\infty} \frac{f^{(k)}(c)}{n!}(x-c)^k. \tag{5.27}$$

The infinite sum (5.27) is called a *Taylor (Maclaurin if* $c = 0$*)* series.

Maclaurin Series for the Basic Transcendental Functions

In this section, we show that the sine, cosine, and exponential functions can be represented by Maclaurin series. We will make use of the following observation

$$\text{For all A>0,} \quad \lim_{n\to\infty} \frac{A^n}{n!} = 0. \tag{5.28}$$

In the case of $\sin x$, we have

$$R_n(x) = (\sin \text{or} \cos)(\xi_n)\frac{x^{n+1}}{(n+1)!}. \tag{5.29}$$

Since the values of both the sine and cosine function lie between -1 and 1, there is the obvious estimate

$$|R_n(x)| \leq \frac{|x|^{n+1}}{(n+1)!}$$

and it follows from (5.28), that $\lim_{n\to\infty} R_n(x) = 0$ in this case too. A similar argument yields the same conclusion for $\cos x$. Making use of (5.21)–(5.23), we obtain the Maclaurin series

$$\begin{aligned}
e^x &= \sum_{k=0}^{\infty} \frac{x^k}{k!} = 1 + x + \frac{x^2}{2!} + \frac{x^3}{3!} + \cdots \\
\sin x &= \sum_{k=0}^{\infty} (-1)^k \frac{x^{2k+1}}{(2k+1)!} = x - \frac{x^3}{3!} + \frac{x^5}{5!} - \cdots \\
\cos x &= \sum_{k=0}^{\infty} (-1)^k \frac{x^{2k}}{(2k)!} = 1 - \frac{x^2}{2!} + \frac{x^4}{4!} - \cdots .
\end{aligned}$$

These Maclaurin series give rise to a remarkable identity of Euler. Substituting the imaginary quantity $i\theta$ for x in the series for e^x gives

$$\begin{aligned}
e^{i\theta} &= i\theta - \frac{\theta^2}{2!} - i\frac{\theta^3}{3!} + \frac{\theta^4}{4!} + i\frac{\theta^5}{5!} - \frac{\theta^6}{6!} - i\frac{\theta^7}{7!} + \cdots \\
&= 1 - \frac{x^2}{2!} + \frac{x^4}{4!} - \frac{x^6}{6!} + \cdots + i(x - \frac{x^3}{3!} + \frac{x^5}{5!} - \frac{x^7}{7!} \cdots) \\
&= \cos\theta + i\sin\theta.
\end{aligned} \tag{5.30}$$

Note that Euler's identity yields a 1 line proof of de Moivre's theorem (Section 4.6):

$$\left(\cos\theta + i\sin\theta\right)^n = \left(e^{i\theta}\right)^n = e^{in\theta} = \cos n\theta + i\sin n\theta.$$

Setting $\theta = \pi$ in (5.30) and recalling that $\cos\pi = -1$ and $\sin\pi = 0$ we obtain *Euler's formula*

$$e^{\pi i} + 1 = 0. \tag{5.31}$$

This equation is said to be the most beautiful formula in mathematics.

Exercise 5.13.* *Use the Maclaurin series*

$$e = 1 + \frac{1}{2!} + \frac{1}{3!} + \cdots$$

to prove that e is irrational. (Hint: Assume that $e = p/q$ is rational. Multiply through by $q!$ and derive a contradiction. If you can complete this argument without further assistance, then you may be a mathematician!)

Finally, we turn to the Maclaurin series for the function $f(x) = \log(1+x)$. This needs a little more work in the estimation of the remainder term, and in order to show convergence to 0 we need to assume that x lies in a restricted range. From (5.24), we obtain

$$R_n(x) = \frac{(-1)^n x^{n+1}}{(1+\xi_n)^{n+1}(n+1)}.$$

Assume $|x| < 1$. Since $|\xi_n| \le |x|$ (recall that $c = 0$ and ξ_n lies between 0 and c),

$$|R_n(x)| \le \frac{|x|^{n+1}}{(1-|\xi_n|)^{n+1}(n+1)} \le \frac{1}{(1-|x|)^{n+1}(n+1)} \to 0 \text{ as } n \to \infty.$$

we obtain

$$\log(1+x) = x - \frac{x^2}{2} + \frac{x^3}{3} - \frac{x^4}{4} + \cdots \quad |x| < 1. \tag{5.32}$$

The series is also valid for $x = 1$ since in this case, $\xi_n > 0$, so we have $|R_n(x)| \le 1/(n+1) \to 0$ as $n \to \infty$. We get the infinite series

$$\log 2 = 1 - \frac{1}{2} + \frac{1}{3} - \frac{1}{4} + \cdots \tag{5.33}$$

There is an easier derivation of (5.32) (although this requires further justification) by performing a term-by-term integration with the geometric series (5.26)

$$\begin{aligned}\log(1+x) &= \int_1^{1+x} \frac{1}{t}\,dt = \int_0^x \frac{1}{1+t}\,dt = \int_0^x \sum_{n=0}^{\infty}(-t)^n\,dt\\ &= \sum_{n=0}^{\infty}\int_0^x (-t)^n\,dt = \sum_{n=0}^{\infty}\frac{(-1)^n}{n+1}.\end{aligned}$$

Suppose finally that a function f is infinitely differentiable at a point c, i.e., the derivatives $f^{(k)}(c)$ exist for all k. Then one may define the Taylor series of f:

$$\sum_{k=1}^{\infty} \frac{f^{(k)}(c)}{n!}(x-c)^k. \tag{5.34}$$

This series may converge only for x in a certain range, as we will see below in the case of the function $\log(1+x)$.This raises the question: at points x where the Taylor series of a function f *does* converge, does it necessarily converge to $f(x)$? The answer to this question is no! The standard example is the function

$$f(x) = \begin{cases} e^{-1/x^2}, & \text{if } x \neq 0, \\ 0, & \text{if } x = 0. \end{cases} \tag{5.35}$$

This function is infinitely differentiable at 0, and $f^{(k)}(0) = 0$, for all k. This is a consequence of the very rapid decay of e^{-1/x^2} to 0, as $x \to 0$.

Exercise 5.14.* *Show that* $f'(0) = 0$.

Hence the Maclaurin series for $f(x)$ is 0. On the other hand, $f(x)$ is non-zero when $x \neq 0$.

Suggestion for Further Reading

James R. Kirkwood. *An Introduction to Analysis.* Chapman and Hall/CRC, 3rd edition, 2021.

CHAPTER 6

Complex Variables

In this chapter, we introduce readers to the field of Complex Variables, the calculus of functions defined using complex numbers. In Sections 6.2 and 6.3, the definitions and results concerning limits and derivatives developed in Chapter 5, are extended to the complex realm. In Section 6.3, a version of integration for complex functions, known as contour integration is introduced, whereby the functions are integrated along curves in the complex plane. Here again, the theory closely parallels real-variable calculus.

The peculiar and striking character of the subject emerges with the appearance of Cauchy's theorem, presented in Section 6.4. Cauchy's theorem spawns a wealth of further results, many of which have no counterparts in the realm of real variable calculus. Some of these results are presented in Section 6.5. One of the consequences of Cauchy's theorem, The Residue Theorem, is explored in Section 6.5.

The Residue Theorem has application in real analysis, making possible the calculation of a variety of integrals and infinite summations that could not otherwise be evaluated. Some examples are given.

6.1 COMPLEX FUNCTIONS

The arithmetic of complex numbers provides for the immediate construction of polynomial and rational functions,

$$P(z) = a_0 + a_1 z + \cdots + a_n z^n$$
$$R(z) = \frac{a_0 + a_1 z + \cdots + a_n z^n}{b_0 + b_1 z + \cdots + b_m z^m}$$

with complex coefficients a_0, b_0, etc.

The introduction of the sine, cosine, and exponential functions into the subject of complex analysis is more problematic. Divorced from their real

DOI: 10.1201/9781003470915-6

foundation, these functions lose their classical meanings. There is no such angle as $3+2i$, so what is does it mean to talk about $\sin(\ 3+2i)$? Or for that matter, to raise the number $2+7i$ to the "power" of $5i$. Real exponential functions occur in mathematical finance, in the context of continuous compound interest. But what does it mean to receive an annual rate of $.02+0.5i$ % on your principal of $10,000+25i$ dollars? Such thoughts must have been in the minds of the mathematicians who first sought to develop a theory of complex functions.

The standard approach is to define these functions by power series (infinite-degree polynomials). To take an example, we saw in Chapter 5 that the sine function can be represented by its Maclaurin series

$$\sin x = \sum_{n=0}^{\infty} \frac{(-1)^n}{2n+1!} x^{2n+1}. \tag{6.1}$$

We will use this series to *define* $\sin z$ for complex z, simply replacing the real argument x in the series by its complex counterpart z. It only makes sense to do this if we know that the series converges for complex z. This issue is addressed by the following theorem.

Theorem 6.1. *(Ratio Test) Consider the power series*

$$\sum_{n=0}^{\infty} a_n (z-c)^n, \ z \in \mathbf{C}. \tag{6.2}$$

Define a number R (called the "radius of convergence" of the series) by

$$R = \lim_{n\to\infty} \left| \frac{a_n}{a_{n+1}} \right|.$$

Then (6.2) converges when $|z-c| < R$ and diverges when $|z-c| > R$.

Applying Theorem 6.1 to the series (6.1), with $c = 0$ and $a_n = (-1)^n/(2n+1)!$, yields [1]

$$\left| \frac{a_n}{a_{n+1}} \right| = \frac{1}{(2n+1)!} \div \frac{1}{(2(n+1)+1)!} = \frac{(2n+3)!}{(2n+1)!} = (2n+3)(2n+2).$$

Hence $R = \lim_{n\to\infty}(2n+3)(2n+2) = \infty$ and we conclude that the Maclaurin series (6.1) for $\sin z$ converges for all $z \in \mathbf{C}$.

The same holds in the case of the Maclaurin series for $\cos z$ and e^z.

Exercise 6.1. *Prove this.*

[1] To be precise, since this series has only odd terms, $a_{2n} = 0$ and $a_{2n+1} = (-1)^n/(2n+1)!$. Hence the Ratio Test implies convergence for $|z^2| < R$, i.e., $|z| < \sqrt{R}$. Since in this case $R = \infty$ it does not make a difference.

We thereby define the exponential, sine, and cosine functions for complex values of z by their Maclaurin series

$$e^z = \sum_{k=0}^{\infty} \frac{z^k}{k!}$$

$$\sin z = \sum_{k=0}^{\infty} (-1)^k \frac{z^{2k+1}}{(2k+1)!}$$

$$\cos z = \sum_{k=0}^{\infty} (-1)^k \frac{z^{2k}}{(2k)!}.$$

Exercise 6.2⋆ *Use this definition of e^z together with the Binomial Theorem, to prove the usual exponential rule*

$$e^{z+w} = e^z \cdot e^w.$$

By contrast with the exponential and trigonometric functions, the Maclaurin series for $\log(1+z)$

$$\sum_{n=1}^{\infty} \frac{(-1)^{n+1}}{n} z^n. \tag{6.3}$$

has a finite radius of convergence given by

$$R = \lim_{n\to\infty} \frac{(n+1)}{n} = 1.$$

Theorem 6.1 thus shows that (6.3) converges for $|z| < 1$ and diverges for $|z| > 1$.[2]

The limited range of convergence suggests that a Maclaurin series might not give a general enough description of the complex logarithm function. For example, the Maclaurin series for $\log(1+z)$ leaves $\log 4$ undefined, since the value $z = 3$ lies outside the range of convergence of the series.

The polar representation suggests another route to defining $\log z$. As inverse function to the exponential e^z, any reasonable choice of natural logarithm ought to satisfy the characteristic log property

$$\log(wz) = \log(w) + \log(z),$$

It would follow that

$$\log(z) = \log(|z|e^{i\arg(z)}) = \log(|z|) + \log(e^{i\arg(z)})$$

$$= \log(|z|) + i\arg(z). \tag{6.4}$$

[2]In general, convergence behavior of the power series on the circle $|z - c| = R$ itself is undetermined by the Ratio Test, and can go either way. For example the series (6.3) happens to converge for $z = 1$ and diverges for $z = -1$.

It is natural to *define* $\log z$ by (6.4). However, the argument of a complex number is not a uniquely defined number (e.g., $\arg i = \pi/2, 5\pi/2, 9\pi/2, \ldots$). It is therefore necessary to designate a range in which to choose $\arg z$, say $[0, 2\pi)$. However, this creates a further a further problem: consider points $z = e^{it}$ lying on the unit circle $\{|z| = 1\}$. As the parameter t increases and gets ever closer to 2π the points z approach 1 (see figure 6.1). At the same time, $\arg z \to 2\pi \neq \arg 1$, since the choice of argument defines $\arg 1$ to have the value 0. So we have a conundrum. If we don't fix a range for $\arg z$ then $\log z$ will have infinitely many values. If we do fix a range, then $\log z$ will be a discontinuous function. Both scenarios are bad in calculus.

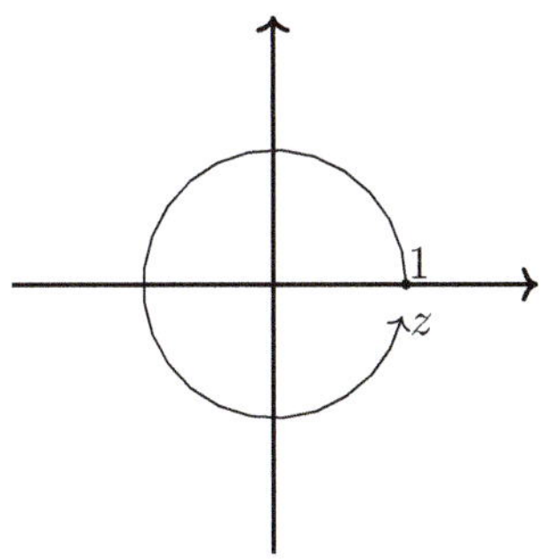

FIGURE 6.1

The standard way to resolve the problem is to remove from the domain of log the negative real axis $(-\infty, 0]$, thereby preventing paths in the domain from winding around 0 (see figure 6.2).

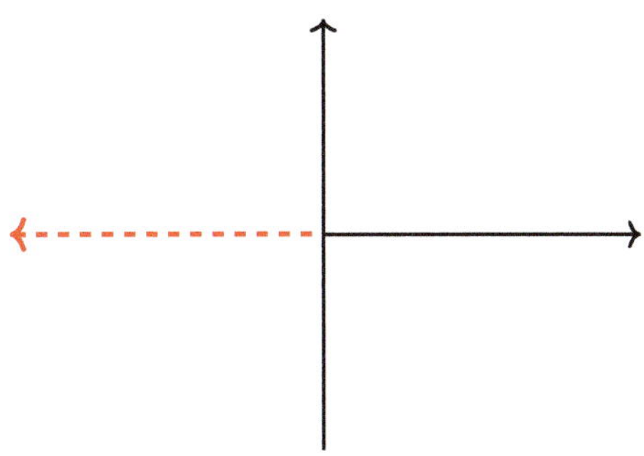

FIGURE 6.2 The cut plane

This results in the standard description of the complex logarithm, as a function defined on the set $\{\mathbf{C} - (-\infty, 0]\}$ by

$$\log(z) = \log(|z|) + i \arg(z), \quad -\pi < \arg(z) < \pi.$$

Note that in the case when z happens to be real-valued and positive, $\log(z)$ by this definition, agrees with the usual real value $\log x$.

Riemann came up with a different, and marvelously original, way to solve the problem of the multi-valued nature of the logarithm. Envisage making a cut in the complex plane along, say, the positive real axis. Take the piece on the left side of the cut and raise it some. At the positive real axis on the raised part, attach a second sheet of exactly the same type. Repeat the process ad infinitum, to create an infinite spiral staircase. A finite part of one is shown in figure 6.3. The resulting configuration (denote it as R) is known as

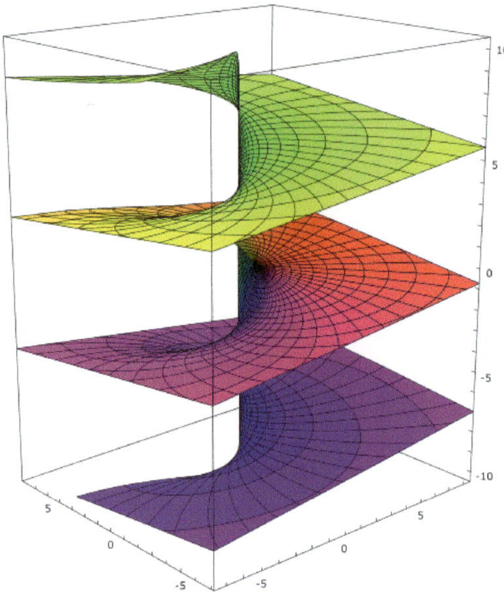

FIGURE 6.3 Riemann surface for the multivalued log function, courtesy of Leonid 2

a *Riemann surface.* Argument, and by association logarithm, are continuously and singly defined functions on R. The reason is, when you have completed a counterclockwise revolution around 0 and your argument has increased by 2π, you are at a *different point* on the surface (one level up).

With this innovation, Riemann introduced a major new tool into complex function theory.

6.2 COMPLEX DIFFERENTIATION

We start this section with some basic definitions. A set $U \subset \mathbf{C}$ is said to be *connected* if any two points in U can be joined by a path lying entirely in U.

We say that U is *open* if, for every point $z \in U$, there is a disc centered at z lying entirely in D.

Exercise 6.3.* *Prove that the disc $D = \{|z| < 1\}$ is open.*

If *a set* is both open and connected then it is said to be a *domain.*

By way of illustration, consider the discs D_1 and D_2 depicted in figure 6.4, where D_2 *includes the boundary circle* and D_1 *does not.* The set D_1 is open: take any point w in D_1 as shown in the figure, then it is possible to draw a small disc centered at w which lies completely inside D_1. By contrast, the disc D_2 is *not open*: take a point w on the boundary of the disc. By definition of D_2, w is a point in the set, yet there is no disc centered at w lying entirely within D_2 (every disc centered at w is bound to go outside of D_2). Both D_1 and D_2 are connected sets, but the set $D = D_1 \cup D_2$ is not connected (clearly, it is not possible to connect a point in D_1 and a point in D_2 by a path lying entirely in D). Because D_1 is both open and connected it is a domain.

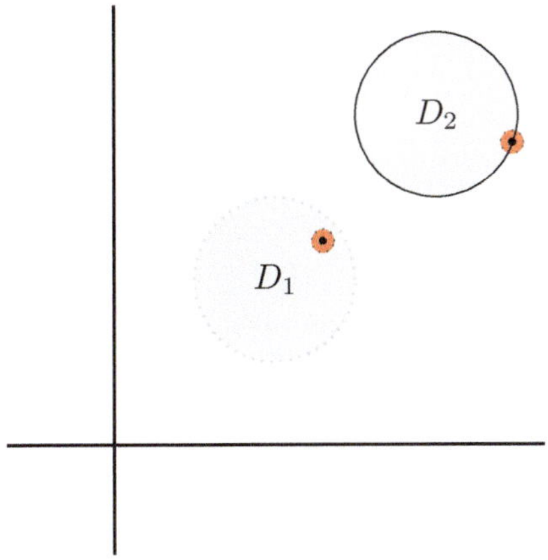

FIGURE 6.4 discs with and without boundary

Suppose that f is a complex complex-valued function defined on a domain $D \subset \mathbf{C}$ (in symbols $f : D \mapsto \mathbf{C}$). We say that f is *(complex) differentiable* if the limit

$$\lim_{h \to 0} \frac{f(z+h) - f(z)}{h} \tag{6.5}$$

exists, in which case the limit is said to be the derivative of f at the point z and denoted $f'(z)$.

A couple of related terms. If f is differentiable on the whole of D, then f is said to be *analytic* on D and if f is defined and analytic on the whole of the complex plane $\mathbf{C}$ then it is said to be an *entire* function.

The Cauchy–Riemann equations

At first glance, the definition of differentiability for complex functions seems the same as for real functions. However, there is a significant difference here. In the real variable case, there are only two ways in which h may approach 0, through values less than 0 and through values larger than 0. In the complex case there are an infinite number of routes for h to get to 0, and the existence of the derivative requires that the limit of the difference quotient be the same in each case! In this sense, complex differentiability is an infinitely more stringent criterion than in the real regime.

To get a feel for what this implies, suppose that f is differentiable at the point $z = x + yi$ with derivative $f'(z) = a + bi$. Taking the limit as $h \to 0$ in (6.5) first through purely *real* values and then through purely *imaginary* values (in the latter case denoting h by ki), we have

$$\begin{aligned} &\lim_{h\to 0} \frac{f(x+yi+h) - f(x+yi)}{h} = \lim_{k\to 0} \frac{f(x+yi+ki) - f(x+yi)}{ki} \\ &= -i \lim_{k\to 0} \frac{f(x+yi+ki) - f(x+yi)}{k} = a + ib. \end{aligned} \tag{6.6}$$

In considering questions of complex differentiability, it is often convenient to express functions f in terms of real-valued functions, called the real and imaginary parts of the function.

$$f(z) = f(x+yi) = U(x,y) + V(x,y)i. \tag{6.7}$$

As an example, consider $f(z) = (x+yi)^2 = (x^2 - y^2) + 2xyi$. Hence in this case $U(x,y) = x^2 - y^2$ and $V(x,y) = 2xy$.

Writing (6.6) in terms of U and V and equating real and imaginary terms on each side of the equations gives

$$\begin{aligned} \lim_{h\to 0} \frac{U(x+h,y) - U(x,y)}{h} &= \lim_{k\to 0} \frac{V(x,y+k) - V(x,y)}{k} = a, \\ \lim_{h\to 0} \frac{V(x+h,y) - V(x,y)}{h} &= -\lim_{k\to 0} \frac{U(x,y+k) - U(x,y)}{k} = b. \end{aligned}$$

The above limits are called partial derivatives.[3] Thus we obtain the partial differential equations

$$\begin{aligned} U_x &= V_y, \\ U_y &= -V_x. \end{aligned} \tag{6.8}$$

[3] For a function of two variables $h(x,y)$, the partial derivative, denoted h_x or $\frac{\partial h}{\partial x}$ is found by fixing y and differentiating with respect to x; similarly h_y.

Equations 6.8 are called the *Cauchy–Riemann equations*. These equations are fundamental to the study of complex functions.

We have shown that the Cauchy–Riemann equations are necessary conditions for the differentiability of a complex function. The following result asserts that the converse holds.

Theorem 6.2. *A function f defined on a domain in the complex plane is differentiable if and only if f satisfies the Cauchy–Riemann equations.*

Theorem 6.2 shows that differentiability is a rather special property of complex functions. To convince yourself of this, think of any two functions U and V of (x, y) and check whether they satisfy (6.8). Probably not, unless you got lucky. Lets try, e.g., $U(x, y) = x^2y^3$ and $V(x, y) = 2xy$. Then $U_x = 2xy^3, U_y = 3x^2y^2, V_x = 2y, V_y = 2x$. The Cauchy–Riemann equations yield

$$\begin{aligned} 2xy^3 &= 2x \\ 3x^2y^2 &= -2y. \end{aligned}$$

Exercise 6.4. *Show that these equations imply $x = y = 0$.*

Hence the corresponding complex function $f(z) = x^2y^3 + 2xyi$ is not differentiable anywhere except (possibly) at the single point $z = 0$.

On the positive side, consider the function e^z, which we can separate into real and imaginary parts using Euler's identity,

$$e^z = e^{x+iy} = e^x(\cos y + i \sin y).$$

to obtain $U(x, y) = e^x \cos y$, $V(x, y) = e^x \sin y$. Since

$$\begin{aligned} U_x &= e^x \cos y = V_y, \\ U_y &= -e^x \sin y = -V_x. \end{aligned}$$

we conclude from Theorem 6.2 that e^z is differentiable everywhere in $\mathbf{C}$ (i.e., entire).

As an immediate corollary of Theorem 6.2, we can show that if f is a purely real-valued analytic function, then f must be constant. (The same conclusion holds if f is purely imaginary.) To see this simply note that since $f(z) = U(x, y)$ and $V(x, y) = 0$, equations 6.8 yield $U_x = U_y = 0$. Hence U is constant.

Exercise 6.5.* *Prove that if an entire function $f = U + Vi$ maps the complex plane into the circle*

$$U^2 + V^2 = r^2,$$

then f is constant.

Complex Differentiable Functions

All the properties of differentiation of real functions set out in Theorem 5.3, product and quotient rules, chain rule, etc., hold in the complex case. In particular, polynomial and rational functions are differentiable, with the expected formulas for the derivatives. An example is

$$\left(\frac{z^2-9}{z^2+4}\right)' = \frac{2z(z^2+4)-(z^2-9)(2z)}{(z^2+4)^2} = \frac{-10z}{(z^2+4)^2}.$$

The following theorem yields a further rich supply of differentiable functions in the form of infinite series.

Theorem 6.3. *Suppose the power series*

$$f(z) = \sum_{n=0}^{\infty} a_n(z-c)^n \tag{6.9}$$

has radius of convergence R. Then for $|z-c| < R$, the function $f(z)$ is differentiable and has derivative

$$f'(z) = \sum_{n=1}^{\infty} na_n(z-c)^{n-1}. \tag{6.10}$$

Furthermore, the derived power series (6.10) has the same radius of convergence R.

Theorem 6.3 can be iterated. Since the radius of the power series (6.9) remains unchanged after differentiating, we may apply the theorem to f' and obtain a power series for f'', etc. Continuing in this fashion, we conclude that f possesses derivatives of all orders within the disc $D = \{x \in \mathbf{C} : |z-c| < r\}$. (We say that f is "infinitely differentiable" on D.)

Exercise 6.6. *Use Theorem 6.3 to prove that*

$$\begin{aligned}
\left(e^z\right)' &= e^z, \\
\sin'(z) &= \cos z, \\
\cos'(z) &= -\sin z.
\end{aligned}$$

6.3 COMPLEX INTEGRATION

Up to this point, the calculus of complex functions closely parallels its real-valued counterpart. There is an entirely analogous theory of limits and differentiation, the usual menagerie of differentiable functions: polynomial, rational, trigonometric, etc., and familiar rules of computation. The unique and remarkable features of the subject appear only when we introduce complex integration. This is the subject of the present section.

Contour Integrals

We say that a curve γ in $\mathbf{C}$ is *smooth* if there exists a parameterization $\gamma = \{\gamma(t), a \le t \le b\}$ such that $\gamma(t)$ is differentiable and $\gamma'(t)$ continuous. Let $f : U \mapsto \mathbf{C}$ be a continuous function and $\gamma \subset U$ be a smooth such curve. Define the integral[4]

$$\int_\gamma f(z)dz = \int_a^b f(\gamma(t))\gamma'(t)dt. \tag{6.11}$$

The second integral in (6.11) can be interpreted in terms of two real integrals. Writing $f = u + vi$ and $\gamma = \alpha + \beta i$ in real and imaginary parts, we define the integral as

$$\int_a^b (u + vi)(\gamma(t))(\alpha + \beta i)'(t)dt =$$

$$\int_a^b \big(u(\gamma(t))\alpha'(t) - v(\gamma(t))\beta'(t)\big)dt + i\int_a^b \big(u(\gamma(t))\beta'(t) + v(\gamma(t))\alpha'(t)\big)dt. \tag{6.12}$$

Let γ be a continuous curve in $\mathbf{C}$ comprised of a finite number of smooth pieces $\gamma_1, \gamma_2, \ldots, \gamma_n$, as depicted in figure 6.5. Such a curve is called a *contour.* (The contour is said to be *closed* if the beginning point of γ coincides with the endpoint, i.e., the curve forms a closed loop, and *simple* if γ does not intersect itself at any intermediate points).

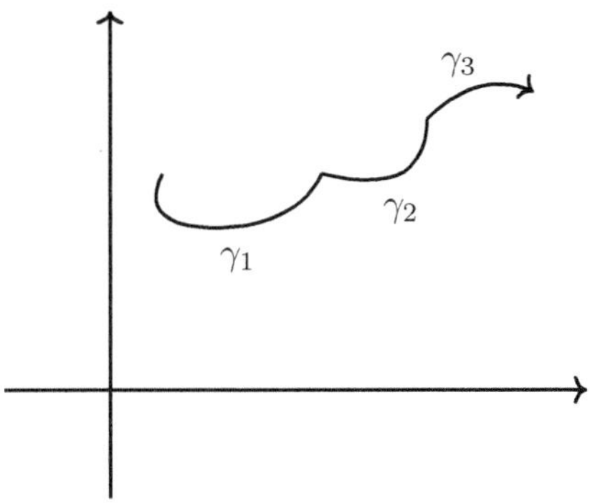

FIGURE 6.5 A contour in the complex plane

Write

$$\gamma = \gamma_1 \cup \gamma_2 \cup \cdots \cup \gamma_n.$$

[4]The parameterization of a curve is not unique. For definition (6.11) to make sense it is necessary to prove that the integral is independent of the choice of parameterization. This is straightforward and relies on the chain rule in (real-variable) calculus. We omit the details.

We define

$$\int_{\gamma} f(z)dz = \sum_{k=1}^{n} \int_{\gamma_k} f(z)dz.$$

Note that contours are oriented. From this point on, we will assume that *all closed contours are oriented in the counterclockwise* direction. For any contour γ, if we denote by $-\gamma$ the contour with the reverse orientation, then it is straightforward to show that

$$\int_{-\gamma} f(z)dz = -\int_{\gamma} f(z)dz. \tag{6.13}$$

There is the following analogue of the First Fundamental Theorem of Calculus for contour integration. It can be proved by appealing to the corresponding result in real-valued calculus.

Theorem 6.4. *Suppose f is a continuous function on a domain D and there exists a function F such that $F' = f$ on D (in this context F is said to be a primitive of f). Then for any contour $\gamma \subset D$ with initial point p and endpoint q,*

$$\int_{\gamma} f(z)dz = F(q) - F(p).$$

In particular, if γ is a closed contour (i.e., $q = p$), then

$$\int_{\gamma} f(z)dz = 0.$$

As an example, let C denote a unit circle of radius r centered at 0. Parameterizing C as $\{re^{it}, 0 \leq t \leq 2\pi\}$ and observing that $(re^{it})' = ire^{it}$, we have

$$\int_C \frac{1}{z}dz = \int_0^{2\pi} \frac{1}{re^{it}} ire^{it}dt = \int_0^{2\pi} idt = 2\pi i. \tag{6.14}$$

Exercise 6.7. *Show that*

$$\int_C \frac{1}{z^n}dz = 0$$

for n an integer greater than or equal to 2.

The salient feature of the functions $1/z$ and $1/z^n, n \geq 2$ as regards contour integration is the singularity that exists at $z = 0$. However, there is an essential difference; the former function lacks a primitive in a neighborhood of the path of integration. (The natural candidate for a primitive of $1/z$, the function $\log z$, is discontinuous on the half-line $(-\infty, 0]$). This problem does not arise for the higher negative powers of z, hence Theorem 6.4 applies to them.

We conclude this section by stating another useful property of contour integrals.

Theorem 6.5.

$$\left|\int_\gamma f(z)dz\right| \leq \max\{|f(z)|, z \in \gamma\} \times l(\gamma),$$

where $l(\gamma)$ denotes the length of the contour γ.

6.4 CAUCHY'S THEOREM

The Cauchy–Goursat theorem, more commonly known simply as Cauchy's theorem is named for two French mathematicians Augustin Louis Cauchy and Édouard Jean-Baptiste Goursat. Cauchy's name is ubiquitous in mathematics; Goursat's much less so. Cauchy's theorem is the central result in the subject of Complex Variables; from this single result spring a host of remarkable properties of analytic functions.

Definition. A region U in the complex plane is said to be *simply connected* if the interior of every closed loop in U is entirely contained within U. In essence, this definition says that the region contains no holes. For example, the annulus (donut-shape) depicted in figure 6.6 is not simply connected. The interior of the red loop contains the central disc, which is not part of the set.

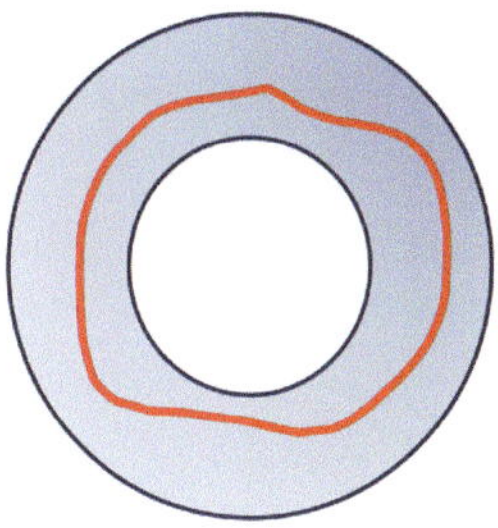

FIGURE 6.6 Annulus with loop

Theorem 6.6. *(Cauchy's theorem) Suppose f is analytic on a simply connected domain D. Then for any simple closed contour C in D*

$$\int_C f(z)dz = 0. \tag{6.15}$$

The theorem has many important consequences, the first of which is a surprising "insensitivity" of contour integrals to the contour of integration, implied by the next theorem.

Theorem 6.7. *Suppose C_1 and C_2 are two simple*[5] *closed contours contained in the domain of f, such that f is analytic on and in the region between the curves. (If this is the case, we say that C_1 and C_2 are homotopic.) Then*

$$\int_{C_1} f(z)dz = \int_{C_2} f(z)dz.$$

To see how useful this theorem is, take any closed contour C containing 0 in its interior. Then we can conclude by Theorem 6.7 and (6.14), that

$$\int_C \frac{dz}{z} = \int_{|z|=1} \frac{dz}{z} = 2\pi i. \tag{6.16}$$

A string of further consequences follow from Cauchy's theorem. In the statements of these theorems we use the following standard notation for the disc and circle of radius r centered at a

$$B_r(a) = \{z \in \mathbf{C} / \ |z-a| < r\},$$
$$C_r(a) = \{z \in \mathbf{C} / \ |z-a| = r\}.$$

Theorem 6.8. *(Cauchy Integral Formula)*

Suppose f is analytic on a domain D and $B_r(a) \subset D$. Then

$$f(a) = \frac{1}{2\pi i} \int_{C_r(a)} \frac{f(z)}{z-a} dz. \tag{6.17}$$

As a consequence of Theorem 6.7, $C_r(a)$ can be replaced by any simple closed contour in $B_r(a)$ which contains z in its interior.

Theorem 6.9. *(Cauchy Integral Formula for Derivatives) Under the assumptions and in the notation of Theorem 6.8, f has derivatives of all orders at a and for all $n \geq 1$*

$$f^{(n)}(a) = \frac{n!}{2\pi i} \int_{C_r(a)} \frac{f(z)}{(z-a)^{n+1}} dz. \tag{6.18}$$

We would be remiss not to point out the truly remarkable nature of this theorem, particularly when viewed from the perspective of real-variable calculus. We started off assuming that f is analytic, i.e., *once* complex differentiable, and ended up with the conclusion that f is *infinitely* differentiable, i.e., has derivatives of all orders! By contrast, consider the function of the real variable x,

$$F(x) = \int_{-1}^{x} |t| dt.$$

---

[5] To say that that a closed curve $C = \{C(t),\ a \leq t \leq b\}$ is *simple* means that the curve has no points of self-intersection except at the endpoints $t = a$ and $t = b$. This would preclude, e.g., the curve $e^{int}, 0 \leq t \leq 2\pi$ for $n > 1$, which describes n counterclockwise loops around 0.

Since the absolute value function is continuous, it follows from FTC(II) that F is differentiable, with derivative $F'(x) = |x|$. But as we have seen, $|x|$ is non-differentiable at 0. Thus F has a first derivative and no higher order derivatives at 0!

Something even stronger than Theorem 6.9 is true.

Theorem 6.10. *Suppose f is analytic on the disc $D \equiv |z-a| \leq r$. Then inside D, f is expressible as its Taylor series centered at a, i.e.,*

$$f(z) = \sum_{n=1}^{\infty} \frac{f^{(k)}(a)}{n!}(x-a)^n. \tag{6.19}$$

We note again the striking difference between Theorem 6.4 and the behavior we saw in Chapter 5, this time in relation to example 5.35.

The case $n = 1$ of Theorem 6.9 has as a corollary the following result, known as Liouville's theorem.

Theorem 6.11. *If f is both entire and bounded, then f is constant.*[6]

Exercise 6.8.⋆ *Prove this.*

Again, we have here a statement that is patently false for real, as opposed to complex, functions, as evidenced by the function $\sin x$ (differentiable everywhere, bounded, and non-constant). The difference is that as a complex function, $\sin z$ is not bounded.

Exercise 6.9. *Prove that $\sin z$ is unbounded on C using the identity*

$$\sin z = \frac{e^{iz} - e^{-iz}}{2i}.$$

A couple of further interesting points about Liouville's theorem:

(1) It can be used it to give a proof by contradiction of the Fundamental Theorem of Algebra (FTA): *Every non-constant polynomial has a complex root.* The argument goes as follows. Suppose to the contrary that $P(z)$ is a non-constant polynomial with no root in $\mathbf{C}$. Then the function $f(z) \equiv 1/P(z)$ is entire. The highest order term in the polynomial dominates for large $|z|$, with the result that

$$|P(z)| \to \infty \text{ as } |z| \to \infty.$$

Hence there exists R such that that $|f(z)| \leq 1$ for $|z| > R$. As a continuous function, f is bounded on the disc $|z| \leq R$, and so bounded on all of $\mathbf{C}$. By Liouville's theorem, f is constant. Hence P is constant, which contradicts our original premise. This contradiction proves FTA.

[6] Recall that "entire" means defined and analytic on the whole of $\mathbf{C}$. To say f is *bounded* means there exists M such that $|f(z)| \leq M$, for all $z \in \mathbf{C}$ (i.e., the range of f is contained in a finite disc).

(2) There is a much stronger version of Liouville's theorem which goes by the name of Picard's theorem:

Theorem 6.12. *If the range of an entire function misses out more than a single value in* $\mathbf{C}$ *then the function is constant.*

As an example, the function e^z never takes the value 0. It follows by Picard's theorem that e^z takes on every other complex value.

6.5 THE RESIDUE THEOREM

Poles and Residues

As an example, consider the function

$$f(z) = \frac{z^2}{(z-2)^2(2z+1)^3}. \tag{6.20}$$

The function has singularities, that is values of z where it is undefined. These occur where the denominator is 0, and this happens when $z = 2$ and $z = -1/2$. These points are said to be *poles* of f.

Suppose we wish to evaluate the integral of f around a simple closed contour γ lying inside the disc $D \equiv \{|z-2| \leq 1\}$. Write

$$f(z) = \frac{1}{(z-2)^2} g(z),$$

where

$$g(z) = \frac{z^2}{(2z+1)^3}$$

and observe that the function g is analytic on the disc $D \equiv \{|z_1| \leq 1\}$. (The only singularity of g occurs $z = -1/2$, which lies outside D. By Theorem 6.10, within D, g has a Taylor series[7] centered at 2,

$$g(z) = a_{-2} + a_{-1}(z-2) + a_0(z-2)^2 + a_1(z-2)^3 + \cdots$$

Hence

$$\begin{aligned} f(z) &= \frac{a_{-2}}{(z-2)^2} + \frac{a_{-1}}{z-2} + a_0 + a_3 1z - 2) + \cdots \\ &= \frac{a_0}{(z-2)^2} + \frac{a_1}{z-2} + h(z), \end{aligned}$$

where h is an analytic function on D.

$$\int_\gamma f(z)dz = \int_\gamma \frac{a_{-2}}{(z-2)^2} dz + \int_\gamma \frac{a_{-1}}{z-2} dz + \int_\gamma h(z)dz.$$

[7]The reason for the strange labeling of the coefficients will very shortly become clear.

The first integral on the right-hand side is 0 by Theorem 6.4 because the integrand has a primitive, and the third is 0 by Cauchy's Theorem 6.6. The second integral is $2\pi i a_{-1}$ by a calculation similar to that of (6.16). Thus we conclude

$$\int_\gamma \frac{z^2\,dz}{(z-2)^2(2z+1)^3} = 2\pi i a_{-1}. \tag{6.21}$$

Definition. Suppose a function f has a series expansion

$$\sum_{k=-n}^{\infty} a_k (z-c)^k$$

in a deleted disc $\{0 < |z-c| < r\}$, for some $r > 0$. Then we say f has a *pole of order* n at c, and that the coefficient a_{-1} in the series is the *residue* of f at c.

The series in the definition is called a *Laurent series*. The residue of f at c is denoted $Res(f, c)$.

Definition. A function whose Laurent series expansion contains infinitely many negative powers of $z - z_0$ is said to have an *essential singularity* at z_0.

An example is the function

$$e^{1/z} = 1 + \frac{1}{z} + \frac{1}{2!z^2} + \frac{1}{3!z^2} + \cdots$$

The behavior of a complex function in a neighborhood of an essential singularity is extremely bizarre. It can be shown that the function takes on every value (with the possible exception of a single value) *infinitely many times*!

Exercise 6.10.* *Use Picard's theorem to show that this is the case for the function* $e^{1/z}$.

Consider again the function (6.20). Suppose now that γ is a simple closed contour contained inside the disc $D = \{|z + 1/2| \leq 1\}$. Arguing as above, we find

$$\int_\gamma \frac{z^2\,dz}{(z-2)^2(2z+1)^3} = 2\pi i Res(f, -1/2). \tag{6.22}$$

Suppose γ is a simple closed contour whose interior includes both poles $-1/2$ and 2. By adding a line to γ as indicated in figure 6.7, we separate γ into the (union of the) two contours γ_1 (rpqr) and γ_2 (sqps). (Note that, the added line is traversed by γ_1 and γ_2 in opposite directions.) By (6.21) and (6.22), we have

$$\begin{aligned}\int_\gamma \frac{z^2\,dz}{(z-2)^2(2z+1)^3} &= \int_{\gamma_1} \frac{z^2\,dz}{(z-2)^2(2z+1)^3} + \int_{\gamma_2} \frac{z^2\,dz}{(z-2)^2(2z+1)^3} \\ &= 2\pi i\big(Res(f, -1/2) + Res(f, 2)\big).\end{aligned}$$

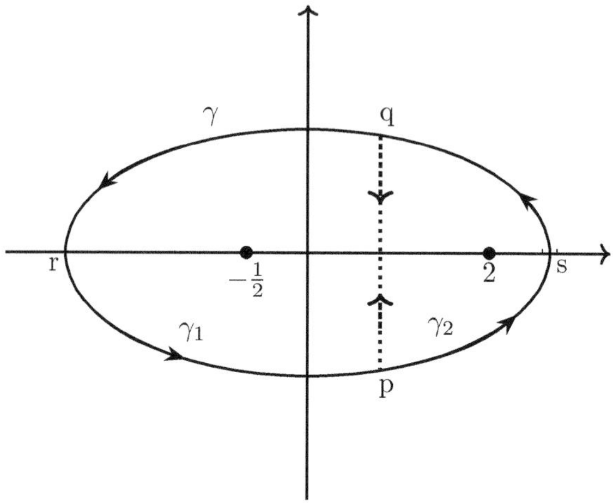

FIGURE 6.7 A contour passing around two poles

The Residue Theorem

The previous example motivates the following major result known as the Residue Theorem.

Theorem 6.13. *Suppose f is analytic on a domain D, except at a finite number of poles. Let γ denote a simple closed contour in D which avoids the poles of f. Then*

$$\int_\gamma f(z)\,dz = 2\pi i \sum_{k=1}^{n} Res(f, c_k), \tag{6.23}$$

where $c_1, c_2, \ldots, c_n$ are the poles of f contained in the interior of γ.

The following theorem gives an efficient way to calculate residues.

Theorem 6.14. *Suppose f has a pole of order n at c. Then*

$$Res(f, c) = \frac{1}{(n-1)!} \lim_{z\to c} \left[(z-c)^n f(z)\right]^{(n-1)}.$$

Exercise 6.11.* *Use the Residue Theorem to compute the contour integral*

$$\int_{|z=3|} \frac{z\,dz}{(z-1)(z-2)^2}.$$

The Residue Theorem is a powerful computational tool in calculus. As an example we are going to use it to compute the integral

$$\int_{-\infty}^{\infty} \frac{e^{itx}}{x^2+1}\,dx. \tag{6.24}$$

This integral arises in probability theory as the characteristic function of the Cauchy distribution (yet another place where Cauchy's name comes up!). Define

$$f(z) = \frac{e^{itz}}{z^2+1}.$$

Because the function e^{itz} is entire, the only poles of f are the points where $z^2+1=0$, i.e., $z=\pm i$. These are poles of order 1 (called simple poles). By Theorem 6.14

$$Res(f,i) = \lim_{z\to i}(z-i)\frac{e^{itz}}{z^2+1}. = \lim_{z\to i}\frac{e^{itz}}{z+i} = \frac{e^{-t}}{2i}. \tag{6.25}$$

For $R>1$, let C_R denote the closed contour consisting of the real interval $[-R,R]$, followed by the semi-circular arc C_R in the upper complex plane, with initial point $(R,0)$ and endpoint $(-R,0)$ (see figure 6.8).

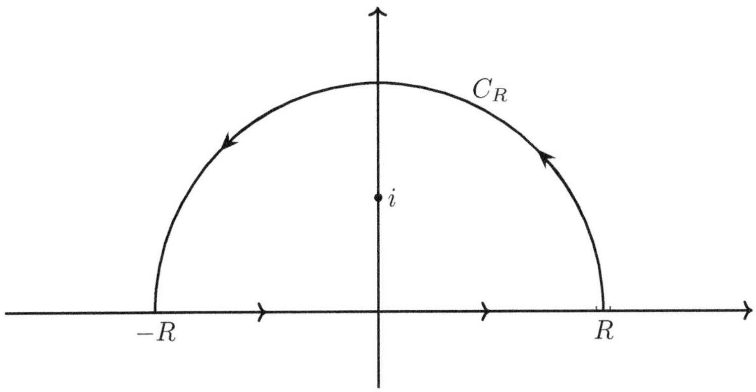

FIGURE 6.8 Semicircular contour

By the Residue Theorem and (6.25)

$$\int_{\gamma_R}\frac{e^{itz}}{z^2+1}dz = \int_{-R}^{R}\frac{e^{itx}}{x^2+1}dx + \int_{C_R}\frac{e^{itz}}{z^2+1}dz = 2\pi i\frac{e^{-t}}{2i} = \pi e^{-t}. \tag{6.26}$$

Note that on C_R, $z=x+yi$ has positive imaginary part y, so

$$\left|e^{itz}\right| = e^{ix-y} = \left|e^{ix}\right|e^{-y} = e^{-y} \le 1.$$

By Theorem 6.5.

$$\left|\int_{C_R}\frac{e^{itz}}{z^2+1}dz\right| \le \text{length }(C_R)\cdot\frac{1}{R^2-1} = \frac{\pi R}{R^2-1} \to 0 \text{ as } r\to\infty.$$

Taking $R\to\infty$ in (6.26), we obtain

$$\int_{-\infty}^{\infty}\frac{e^{itx}}{x^2+1}dx = \pi e^{-t}.$$

The Residue Theorem and the Basel Problem

The Residue Theorem can also be used to compute the sum of a wide collection of infinite series. An example is

$$\sum_{k=1}^{\infty} \frac{1}{k^2} = \frac{\pi^2}{6}. \tag{6.27}$$

The evaluation of this infinite sum was posed as a challenge to mathematicians by Pietro Mengoli in 1650, later named the Basel Problem. Its solution by Euler in 1734 was a major achievement and brought him instant fame at the age of twenty eight. Nowadays there are many different ways to compute (6.27). We are going to give a derivation based on the Residue Theorem.

The idea is to integrate the function

$$f(z) = \frac{\cot(\pi z)}{z^2} = \frac{\cos(\pi z)}{z^2 \sin(\pi z)}$$

over the square contour C_N depicted in figure 6.9. The rationale for choosing the function and the contour in this way is threefold:

(i) f has poles at the points $k\pi, k = 1, 2, 3 \ldots.$

(ii) The contour C_N avoids these poles.

(iii) C_N becomes infinite in both directions as $N \to \infty$. This (along with the term z^2 in the denominator of f) permits an estimate of the integral which shows that it vanishes in the limit, and therefore does not need to be explicitly evaluated.

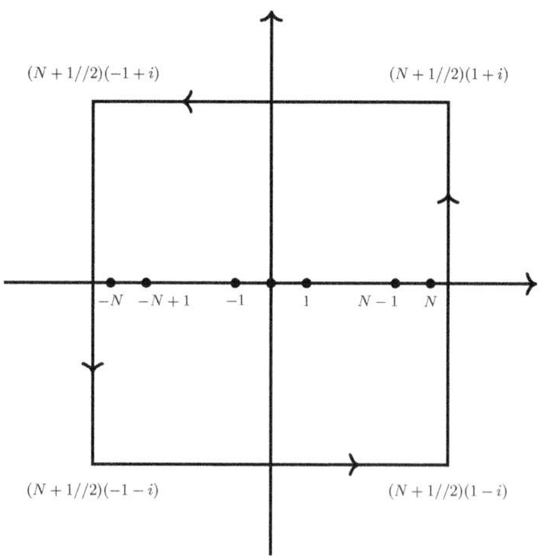

FIGURE 6.9 The contour C_N

The function f has a pole of order 3 at $z = 0$. According to Theorem 6.14

$$Res(f, 0) = \lim_{z\to 0} \left(z \cot(\pi z)\right)'''.$$

However, the third-order derivative is a little cumbersome and it is easier to compute the residue using the Maclaurin series for $\sin z$ and $\cos z$ and the geometric series:

$$\begin{aligned}\frac{\cot(\pi z)}{z^2} &= \frac{1}{z^2}\Big(1 - \frac{(\pi z)^2}{2} + \frac{(\pi z)^4}{24} + \cdots\Big)\Big((\pi z) - \frac{(\pi z)^3}{6} + \frac{(\pi z)^5}{120} + \cdots\Big)^{-1} \\ &= \frac{1}{\pi z^3}\Big(1 - \frac{(\pi z)2}{2} + \frac{(\pi z)^4}{24} + \cdots\Big)\Big(1 - \Big[\frac{(\pi z)^2}{6} - \frac{(\pi z)^4}{120} + \cdots\Big]\Big)^{-1} \\ &= \frac{1}{\pi z^3}\Big(1 - \frac{(\pi z)^2}{2} + \cdots\Big)\Big(1 + \frac{(\pi z)^2}{6} + \cdots\Big)\end{aligned}$$

To determine the residue at 0 (i.e., the coefficient of z^{-1} in the above expression) it is only necessary to pick out the coefficient of z^2 in the product of the brackets, and this is

$$-\frac{\pi^2}{2} + \frac{\pi^2}{6} = -\frac{\pi^2}{3}.$$

This results in

$$Res(f, 0) = -\frac{\pi}{3}.$$

The other poles of f inside C_N are simple poles and occur at $z = \pm 1, \pm 2, \ldots, \pm N$. A similar calculation to the one above (or an application of L'Hôpital's rule) results in

$$Res(f, k) = \lim_{z\to k} (z - k) \cdot \frac{\cot(\pi z)}{z^2} = \frac{1}{\pi k^2}.$$

The Residue Theorem thus gives

$$\int_{C_N} \frac{\cot(\pi z)}{z^2}\, dz = 2\pi i\Big(-\frac{\pi}{3} + \sum_{k=-n, k\neq 0}^{n} \frac{1}{\pi k^2}\Big) = 2\pi i\Big(-\frac{\pi}{3} + \frac{2}{\pi}\sum_{k=1}^{n} \frac{1}{k^2}\Big). \tag{6.28}$$

We now show that the function

$$\cot(\pi z) = i \cdot \frac{e^{i\pi z} + e^{-i\pi z}}{e^{i\pi z} - e^{i\pi z}}$$

is bounded on the contour C_N.

Firstly, note that on the bottom side of the rectangle, $z = tx - (N + 1/2)i$, so

$$|\cot(\pi z)| = \left|\frac{e^{\pi xi}e^{-\pi(N+1/2)} + e^{-\pi xi}e^{\pi(N+1/2)}}{e^{\pi xi}e^{-\pi(N+1/2)} - e^{-\pi xi}e^{\pi(N+1/2)}}\right|.$$

Using the inequalities $|w + z| \leq |w| + |z|$ and $|z - w| \geq |z| - |w|$, together with the fact that $|e^{\pi xi}| = 1$, we have

$$|\cot(\pi z)| \leq \left|\frac{e^{\pi(N+1/2)} + e^{-\pi(N+1/2)}}{e^{\pi(N+1/2)} - e^{-\pi(N+1/2)}}\right| = \left|\frac{1 + e^{-\pi(2N+1)}}{1 - e^{-\pi(2N+1)}}\right| < 3. \tag{6.29}$$

Exercise 6.12. *Prove the second inequality in (6.29).*

On the right side of the rectangle C_N, z has the form $(N + 1/2) + iy$. This implies

$$e^{i\pi z} = e^{-\pi y} e^{\pi(N+1/2)i} = e^{-\pi y}\big(\cos(\pi(N+1/2)) + i\sin(\pi(N+1/2))\big) = \pm i e^{-\pi y}$$

and

$$e^{-i\pi z} = \mp i e^{-\pi y}$$

with the signs depending on whether N is even or odd. Hence, in this case

$$|\cot(\pi z)| \leq \left| \frac{e^{\pi} y - e^{-\pi y}}{e^{\pi y} + e^{-\pi y}} \right| \leq 1.$$

Similar estimates hold on the two remaining sides of C_N.

Theorem 6.5 now yields

$$\left| \int_{C_N} \frac{\cot \pi z}{z^2} \right| \leq \text{length } (C_N) \cdot \max_{z \in C_N} \left| \frac{\cot \pi z}{z^2} \right| \leq \frac{3(4N + 2)}{N^2} \to 0 \text{ as } N \to \infty.$$

Taking $N \to \infty$ in (6.28) gives

$$-\frac{\pi}{3} + \frac{2}{\pi} \sum_{k=1}^{\infty} \frac{1}{k^2} = 0.$$

Hence

$$\sum_{k=1}^{\infty} \frac{1}{k^2} = \frac{\pi^2}{6}. \tag{6.30}$$

Exercise 6.13.* *Use (6.30) to evaluate the sum*

$$1 + \frac{1}{3^2} + \frac{1}{5^2} + \frac{1}{7^2} + \cdots$$

Suggestion for Further Reading

Theodore W. Gamelin. *Complex Analysis.* Springer, 2021.

CHAPTER 7

Graph Theory

We start this chapter with the bridges of Königsberg problem. Euler's description of the problem in terms of edges and vertices is the foundational example in graph theory, showing how simple diagrams can contain all the essential details of certain problems.

After looking at basic properties that apply to all graphs, we turn attention to graphs that can be drawn in the plane (or on the sphere) without edges crossing. These graphs can be characterized by the *Euler characteristic*, a number calculated by counting the number of vertices, edges and faces. We show that the regular polygons considered at the end of the first chapter all have Euler characteristic of 2.

Next we consider graphs drawn on other surfaces. We see that the Euler characteristic of the graph, if drawn in an appropriate way, gives a number that characterizes the surface.

We then look at graph coloring and the four color problem. The proof of the four color problem is the first major mathematical proof using a computer, and so beyond the scope of this book. But we do give a proof of the five color problem.

We briefly study Eulerian circuits and Hamiltonian cycles—ways of traveling along the edges to every vertex. Here, we give Euler's solution to the bridges of Königsberg problem. We see that two seemingly similar problems are really quite different. This observation is important for computer science and is discussed further in the last chapter.

As we mentioned, the Euler characteristic tells us about the surface on which the graph is drawn. We conclude the chapter with a remarkable theorem showing how the intrinsic geometry of a surface is linked to this number.

DOI: 10.1201/9781003470915-7

7.1 HISTORY

Graph theory is generally considered to begin with Euler and the bridges of Königsberg problem. The Pregel River ran through Königsberg (now Kaliningrad, Russia). In the river were two islands. Seven bridges connected the islands to one another and to the mainland. Figure 7.1 shows a map of the city as it was in 1736. It depicts all of the smaller island, but only a portion of the other island to its right. The smaller island has five bridges: two connect it to the northern bank of the river; two to the southern bank; the other bridge connects to the larger island. There are two other bridges on the larger island connecting it to both banks of the river.

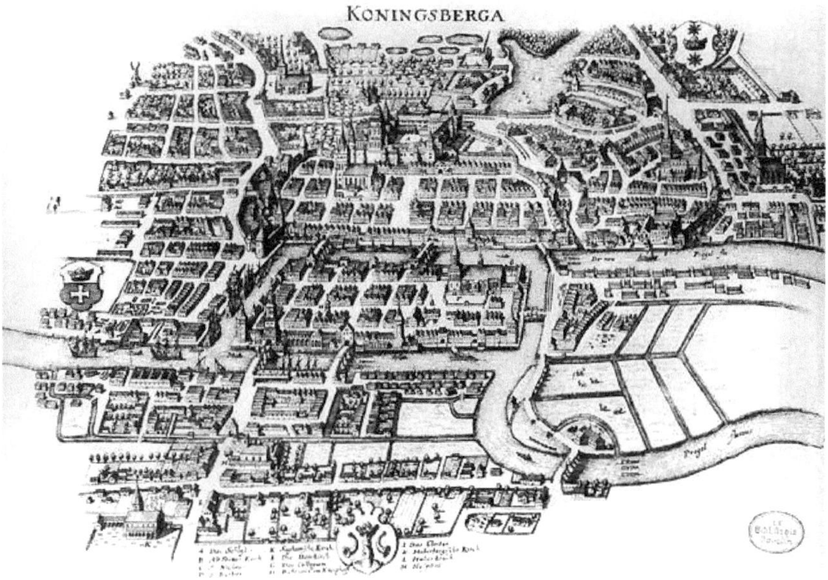

FIGURE 7.1 Map of Königsberg

The Königsberg bridge problem was to devise a walk that crossed each bridge once and only once. There are four regions of land: the two islands, the city north of the river, and the city south of the river. Euler realized he could depict each of the regions as dots and the bridges as lines connecting the dots as shown in figure 7.2.

In 1736, Euler showed that there was no solution to the problem. We will see his solution later.

Certain problems like this allow us to abstract the pertinent features to form a model consisting of dots and lines connecting them. James Joseph Sylvester coined the term "graph" for these networks. In 1891, Julius Petersen wrote an article where instead of using them as a tool for solving some other problem, he studied them as abstract objects in their own right. Ever since

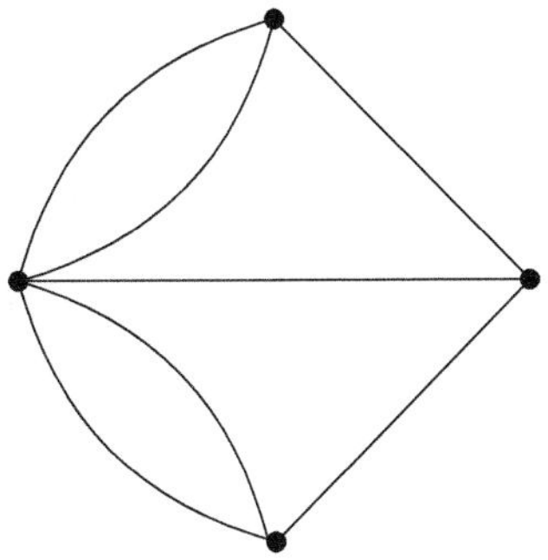

FIGURE 7.2 Simplified map of Königsberg bridges

this area of mathematics has been known as *graph theory*. Instead of "dots" and "lines" the terms "vertices" and "edges" are used.

A *graph* consists of a finite set of *vertices* V together with a finite set of *edges* E. Each edge connects two vertices. There is at most one edge between any two vertices.

Example 7.1. *We will let* $V = \{a, b, c, d\}$. *The set* E *can be defined by the vertices the edges connect. For our example, we let* $E = \{\{a, b\}, \{a, c\}, \{b, c\}\}$. *The graph is depicted in figure 7.3.*

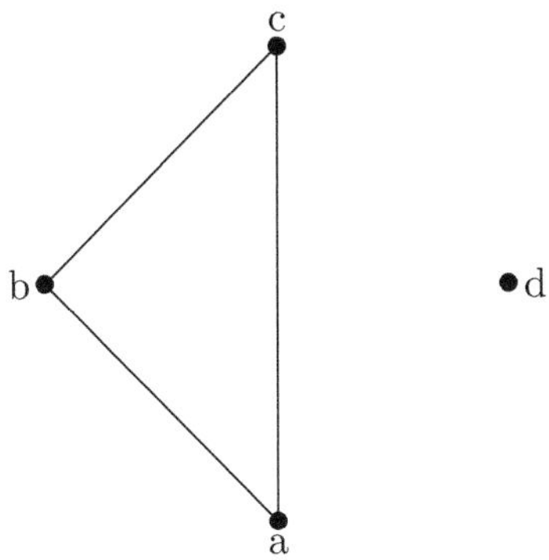

FIGURE 7.3 Graph for example 7.1

Figure 7.2 has two edges going from the bottom vertex to the one on the left. This is not allowed for a graph—there can be at most one edge between any two vertices. If we allow more than one edge between two vertices we call the structure a *multigraph*.

Multigraphs

Multigraphs are studied as a part of graph theory. We will look at them later when we return to the Königsberg problem.

FIGURE 7.4 Leonhard Euler, portrait by Jacob Emanuel Handmann, 1753

Given a graph, we will let v denote the number of vertices and e the number of edges. The *degree* of a vertex is the number of edges that attach to it. So in figure 7.3 vertices a, b, and c have degree 2 and vertex d has degree 0. It is straightforward to prove the following.

Theorem 7.1. *In any graph the sum of the degrees of the vertices equals twice the number of edges.*

Exercise 7.1.* *Prove Theorem 7.1.*

We obtain the following corollary as an immediate consequence of Theorem 7.1

Corollary 7.1. *In any graph there are an even number of vertices of odd degree.*

A standard example to illustrate these results is people shaking hands at a party. The people attending the party correspond to the vertices and the handshakes to edges. Theorem 7.1 tells us that the total number of handshakes is even. Corollary 7.1 says there are an even number of people who shake hands an odd number of times.

If a graph has v vertices and one vertex has edges connecting it to every other vertex in the graph, it will have degree $v - 1$. If a vertex is isolated, has no edges connected to it, it will have degree 0. The maximum degree a vertex can have is $v - 1$, and the minimum is 0. This gives us v possibilities for degrees of vertices, but notice that a graph cannot have both a vertex of degree 0 and of degree $v - 1$. Since we have v vertices and we know it cannot have one of each degree from 0 to $v - 1$, there must be two vertices that have the same degree.

Theorem 7.2. *In any graph there will be two vertices that have the same degree.*

This tells us that two people at our party shook the same number of hands.

We have shown that there must be at least two vertices with the same degree, but it is possible that every vertex in a graph has the same degree. If all the vertices have the same degree, the graph is called *regular*. Sometimes the degree is attached, so r-regular is a regular graph in which every vertex has degree r.

Exercise 7.2.[*] *Show that if a graph is 1-regular then it must have an even number of vertices.*

A graph is called *complete* if every pair of vertices is connected by an edge. The complete graph with v vertices is denoted by K_v. K_1 consists of a single vertex; K_2 has two vertices with an edge connecting them; K_3 is a triangle with three edges and three vertices; K_4 is depicted in figure 7.5.

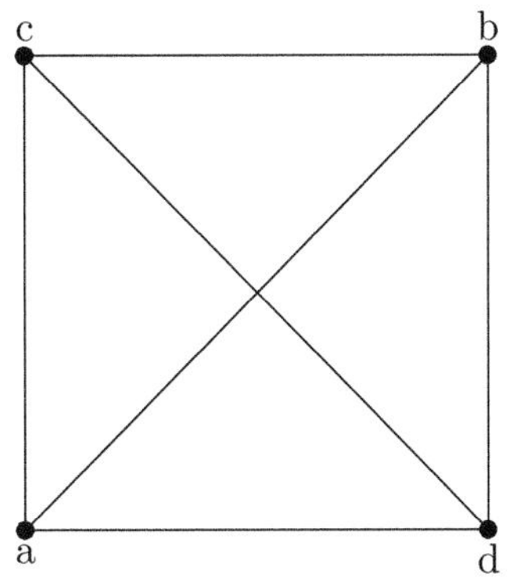

FIGURE 7.5 Complete graph with four vertices—K_4

Exercise 7.3. *Draw K_5. This graph is important later.*

7.2 PLANAR GRAPHS

A graph is *planar* if it can be drawn in the plane (on a page) without any of the edges crossing. If we can draw a graph in the plane and no edges cross, then it is planar. However, if we have an example of a graph which has edges crossing, it is not necessarily non-planar. For example, figure 7.6 shows a graph with four vertices and two edges. The graph is drawn with two edges crossing, but this graph is planar because it is possible to draw it without edges crossing. Figure 7.7 shows one way of redrawing the graph.

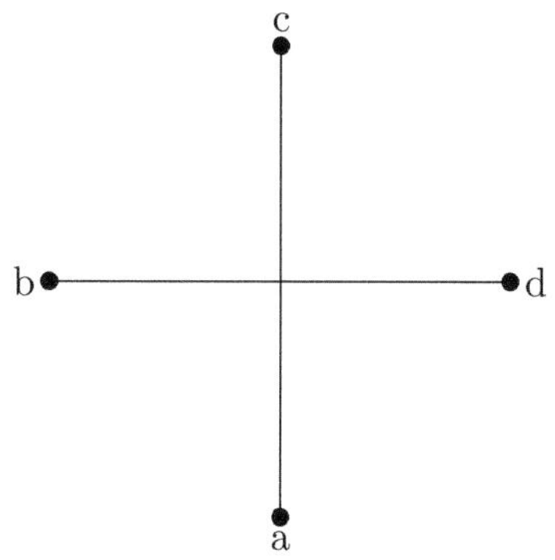

FIGURE 7.6 Graph with four vertices and two edges

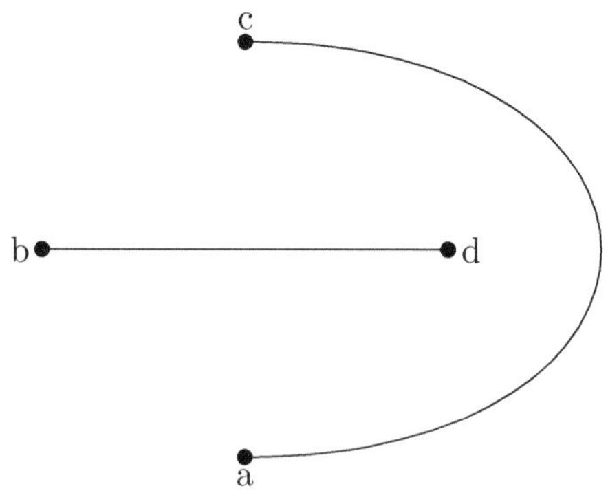

FIGURE 7.7 Graph with four vertices and two edges redrawn

Exercise 7.4.[*] *Show K_4 is planar.*

K_5 is nonplanar

We will show that K_5 is nonplanar using a proof by contradiction. Let a, b, c, d, and e denote the five vertices. We suppose someone has presented us with a diagram of K_5 with no edges crossing.

Since K_5 is complete there are edges from a to b, from b to c, from c to d, from d to e, and from e to a. These five edges form a loop. This loop divides the plane into two regions: the finite one inside, and the infinite one outside. Figure 7.8 shows the vertices with the loop.

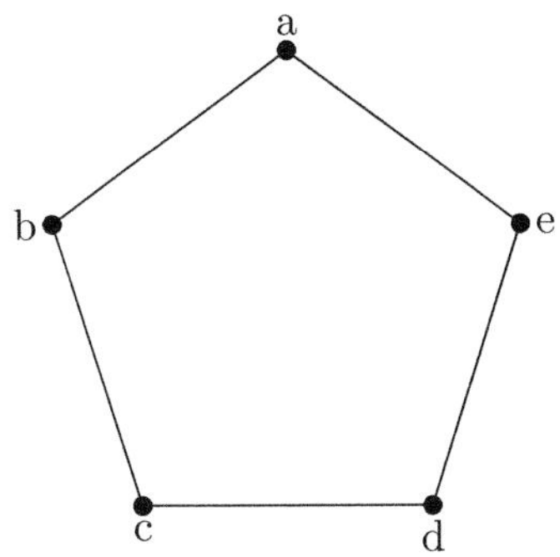

FIGURE 7.8 Portion of K_5 with loop

We have drawn five of the edges. There are another five edges that need to be included: $\{a, c\}, \{a, d\}, \{b, d\}, \{b, e\}$, and $\{c, e\}$. Each of these edges must lie entirely inside our loop or entirely outside. There are two possibilities: either at least three of these edges lie inside, or at least three edges lie outside.

Exercise 7.5. *Show that however you pick three edges from* $\{a, c\}, \{a, d\}$, $\{b, d\}, \{b, e\}$, *and* $\{c, e\}$ *there will always be two that share a common vertex.*

First, we consider the case when three of the edges lie inside the loop. The previous exercise shows that two of these edges share a common vertex. We will assume this vertex is a but exactly the same argument works for the other vertices. Figure 7.9 shows the graph with edges $\{a, c\}$ and $\{a, d\}$ added. But now it is impossible to add the third edge and stay inside the original loop. We conclude it is impossible to have three non-crossing edges inside the loop.

We now consider the case when at least three of the edges are outside the loop. As before, we know that two of them must share a common vertex. Again, we will assume it is a, but exactly the same argument works if you choose any of the other vertices. Figure 7.10 shows the graph with edges $\{a, c\}$ and $\{a, d\}$ added. But now it is impossible to add the third edge and stay outside the original loop. We conclude it is impossible to have three non-crossing edges outside the loop.

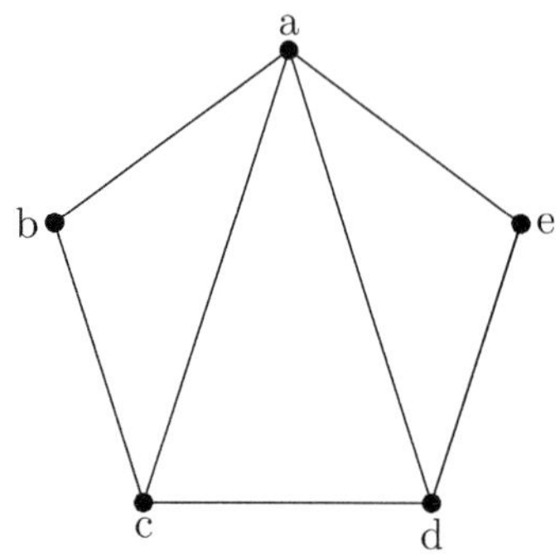

FIGURE 7.9 Portion of K_5 with more inside edges added

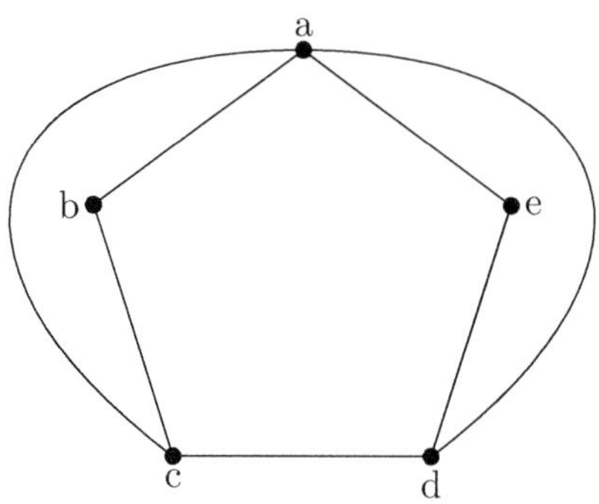

FIGURE 7.10 Portion of K_5 with more inside edges added

We argued that if K_5 was planar, then there has to be a drawing of the graph in the plane with no edges crossing. This lead to the construction of a loop and the observation that there must be at least three edges in the interior or at least three edges outside the loop. We have shown neither of these cases are possible. We conclude that no such drawing can exist. This proves K_5 is nonplanar.

Theorem 7.3. *K_5 is nonplanar.*

The three houses and three utilities problem

Suppose we have three utility companies and three houses. We want to connect each of the three utilities to each of the three houses. We will denote the utility companies by E (electricity), W (water), and G (gas), and the houses by A, B, and C. Figure 7.11 shows a graph with the utilities, houses and connections. The problem is whether this graph can be redrawn without any of the edges crossing. Is this graph planar?

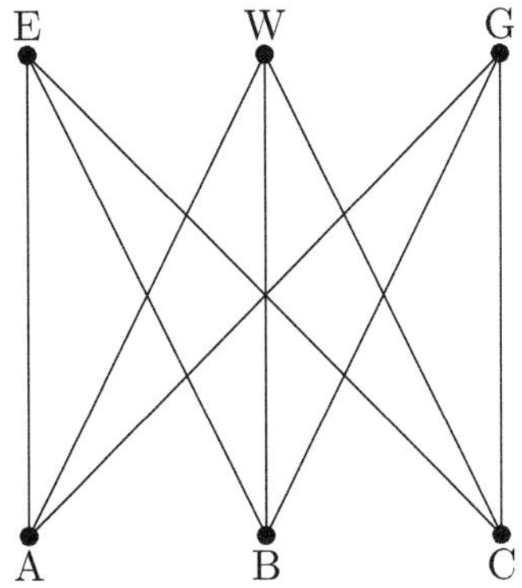

FIGURE 7.11 Three utilities and three houses

This graph is an example of a *bipartite* graph—a graph in which the vertices can be divided into two disjoint sets V_1 and V_2 such that every edge connects a vertex in V_1 to V_2. (There are no edges between vertices in V_1, and no edges between edges in V_2.) For our utility graph the sets of vertices are $\{E, W, G\}$ and $\{A, B, C\}$.

If every vertex in V_1 connects to every vertex in V_2, then we call it a *complete bipartite* graph. The utility graph is a complete bipartite graph.

A complete bipartite graph is denoted by $K_{r,s}$, where r is the number of vertices in V_1 and s is the number of vertices in V_2. Our utility graph is $K_{3,3}$.

Exercise 7.6:[⋆]

Show $K_{2,3}$ is planar.

$K_{3,3}$ is Nonplanar

We will show that $K_{3,3}$ is nonplanar, and consequently there is no solution to the utility problem in the plane with edges that don't cross. Our proof will follow along the lines of the proof that K_5 is nonplanar.

We first construct a loop containing all the vertices. We know that every utility is connected to every house, so there must be the following edges: {A,E}, {E,B}, {B, W}, {W, C}, {C,G}, and {G,A}. If $K_{3,3}$ is planar there is a graph in the plane with no edges that cross. Figure 7.12 depicts this loop.

We have drawn six of the edges. The missing three are: {A,W}, {B,G}, and {C,E}. There are two cases to consider: either at least two of these edges lie inside the loop or at least two edges lie outside. We will show neither is possible.

If we draw one of these edges inside the loop, it divides the interior of the loop into two regions. Both the other two edges have one vertex on one side of this new edge and the other vertex on the other side. It is impossible to draw either of these edges inside the region without crossing the new edge.

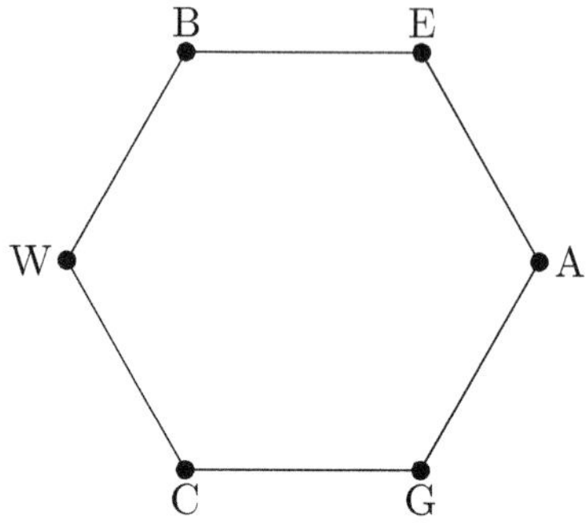

FIGURE 7.12 Portion of $K_{3,3}$ with loop

We conclude there can be at most one of {A,W}, {B,G}, and {C,E} inside the loop. (Figure 7.13 shows the loop with one interior edge.)

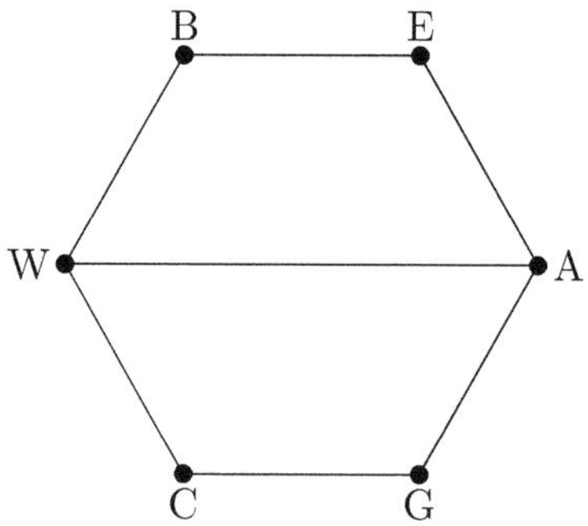

FIGURE 7.13 Portion of $K_{3,3}$ with loop and one interior edge

If we draw one of the edges outside the loop, it divides the exterior of the loop into two regions: one has finite area the other is infinite. It is impossible to draw either of the other two edges outside the original loop without crossing the new edge. We conclude there can be at most one of {A,W}, {B,G}, and {C,E} outside the loop. (Figure 7.14 shows the loop with one exterior edge.)

We argued that if $K_{3,3}$ was planar, then there has to be a drawing of the graph in the plane with no edges crossing. This lead to the construction of a loop and the observation that there must be at least two edges in the interior or at least two edges outside the loop. We have shown neither of these cases are possible. We conclude that no such drawing can exist. This proves $K_{3,3}$ is nonplanar.

Theorem 7.4. *$K_{3,3}$ is nonplanar.*

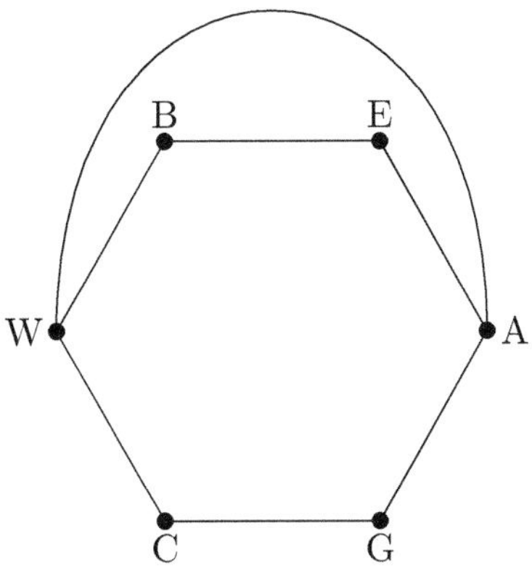

FIGURE 7.14 Portion of $K_{3,3}$ with loop and one exterior edge

Kuratowski's Theorem

A graph H is a *subdivision* of a graph G if either $H = G$ or H can be obtained from G by inserting new vertices into the edges. If we add a vertex to an edge, the original edge is 'chopped' to form two edges in the new graph. All the new vertices have degree 2. Figure 7.15 shows a graph G on the left and a subdivision H on the right. In this graph, the edge from a to c has two new vertices added. The original edge in G becomes three edges in H,

Adding vertices does not affect whether a graph is or is not planar. All graphs that are subdivisions of K_5 are nonplanar and all graphs that are subdivisions of $K_{3,3}$ are nonplanar. It might seem that subdivisions of K_5 and $K_{3,3}$ are very special types of nonplanar graphs, but it turns out that if a graph is nonplanar then one of these graphs is contained within it.

A graph H is a *subgraph* of a graph G if either $H = G$ or H can be obtained from G by removing edges and vertices. Edges must end in vertices, so when a vertex is removed all the edges that connect to that vertex must also be removed. Figure 7.16 shows a graph G on the left and a subgraph H on the right.

The Polish mathematician Kazimierz Kuratowski published the following remarkable theorem in 1930. This result was also proved independently by Orrin Frank and Paul Smith around the same time, but Kuratowski was first to publish, and the theorem is now named after him.

Theorem 7.5. *A graph is planar if and only if it does not contain a subdivision of K_5 or $K_{3,3}$ as a subgraph.*

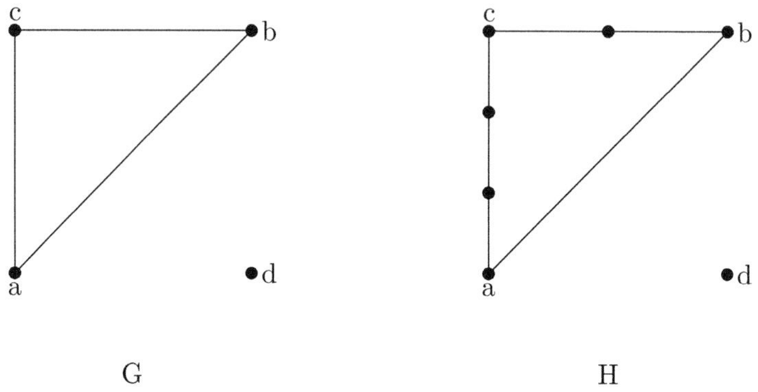

FIGURE 7.15 A graph G with a subdivision H

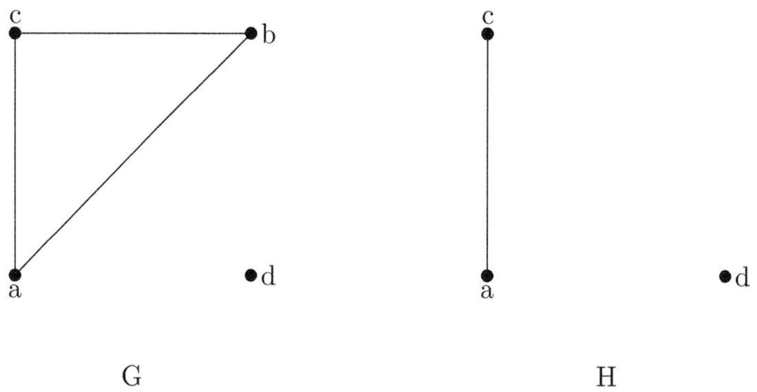

FIGURE 7.16 A graph G with a subgraph H

7.3 THE EULER CHARACTERISTIC

Table 7.1 lists the five Platonic solids. For each solid, the number of vertices, edges and faces are listed.

We will let v, e, and f stand for the number of vertices, edges, and faces. Euler noticed that for each of the solids, the sum $v - e + f$ always equals 2. This number is called the *Euler characteristic*

We can represent each of these solids as graphs in the plane by removing one face and then stretching the remaining object until it is planar. Figure 7.17 shows three of these graphs. We leave the other two as an exercise for the interested reader.

The enclosed regions are called *faces*. We also call the infinite region around the graph a face, sometimes it is called the infinite face. It corresponds to the face we removed from the solid before stretching it. Counting the infinite face as a face means our relationship $v - e + f = 2$ is preserved.

TABLE 7.1 The five Platonic solids with numbers of vertices, edges, and faces

Platonic solid	Vertices	Edges	Faces	$v - e + f$
Tetrahedron	4	6	4	2
Cube	8	12	6	2
Octahedron	6	12	8	2
Dodecahedron	20	30	12	2
Icosahedron	12	30	20	2

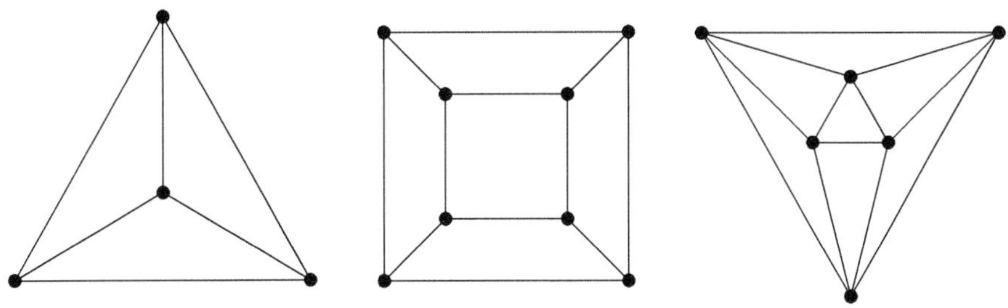

Tetrahedron Cube Octahedron

FIGURE 7.17 Planar graphs of three Platonic solids

We will show that this equation holds for all connected planar graphs. *Connected* means that given any two vertices in the graph there is a path along edges that connects them.

Exercise 7.7* *Show that for any* $m > 2$ *there is a disconnected planar graph with* $v - e + f = m$.

Trees

Every planar graph has the infinite face, so the number of faces must be at least one. Connected planar graphs that only have one face—the infinite face—are called *trees*. Figure 7.18 shows three trees that each have five vertices.

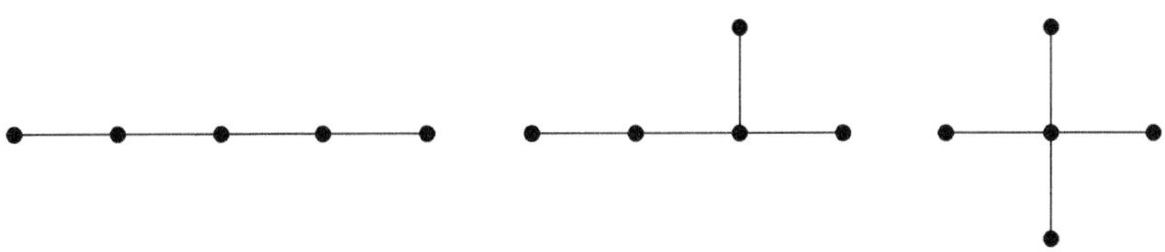

FIGURE 7.18 Three trees with five vertices

The degrees of the vertices are either 1 or 2. The edges that have a vertex of degree 1 are called *leaves*. Every tree with three or more vertices must have at least two leaves. Figure 7.18 shows trees with 2,3 and 4 leaves.

Theorem 7.6. *Every tree with v vertices has $v - 1$ edges.*

Proof. We will use a proof by induction on the number of vertices. The tree with 1 vertex has no edges, and so the theorem is true in this case. The tree with two vertices has one edge between them. Again, the tree is true for $v = 2$.

Now suppose that we have proved the theorem for all trees with n or fewer vertices. We will show it must hold for all trees with $n + 1$ vertices.

Suppose we have a tree with $n + 1$ vertices, then it has at least two leaves. Pick one of them and remove this edge. Also remove the degree one vertex that is attached to this edge. We now have a tree with n vertices, by our inductive hypotheses we know this tree has $n - 1$ edges. To get our original tree we must add back one edge and one vertex, giving $n + 1$ vertices and n edges.

□

Corollary 7.2. *For any tree, $v - e + f = 2$.*

We know that trees only have the infinite face, so $f = 1$. We obtain

$$v - e + f = v - (v - 1) + 1 = 2.$$

The Euler Characteristic for Planar Graphs

We now prove that the Euler characteristic of any connected planar graph is 2.

Theorem 7.7. *Given any connected planar graph with v vertices, e edges and f faces, $v - e + f = 2$.*

Proof. We know the result is true for trees, so we need only consider the cases where the number of faces is 2 or greater. We will use induction on the number of edges. Connected graphs with 0, 1 or two edges are trees. We begin our induction with $e = 3$. There is one connected graph with $e = 3$ and $f = 2$. This is a triangle with three vertices, three edges and two faces. This example has $v - e + f = 2$.

Suppose that we have proved the formula for all connected planar graphs with two or more faces that have n or fewer edges. Suppose that we are given a connected planar graph with two or more faces having $n + 1$ edges. We will show this graph must have Euler characteristic 2.

Given the graph, we know there are two or more faces. There must be a finite face. The edges that form the boundary of this face have the finite face on one side and another face on the other side (possibly the infinite face). If we remove one of these edges the two faces that we separated by the edge become connected. This results in a graph that has one fewer faces and one fewer edge.

This new graph with the removed edge has n edges. If the new graph is a tree, we know it has Euler characteristic 2. If the new graph is not a tree, then by the inductive hypothesis it has Euler characteristic 2.

If we add the edge back to get the graph with $n + 1$ edges we increase e and f by 1, but v is unchanged. This means the Euler characteristic is still 2. □

Each edge in a planar graph has a face on either side. It is possible for an edge to have the same face on both sides. Figure 7.19 shows a planar graph with the edges labeled. Edges E_5 and E_3 have one face on both sides. All the other edges have the finite face on one side and the infinite face on the other. The *boundary* of a face consists of all the edges that have that face on one or both sides. So, in the example, the boundary of the finite face consists of the five edges E_1, E_2, E_4, E_6, and E_5. The boundary of the infinite face has five edges—all the edges except for E_5.

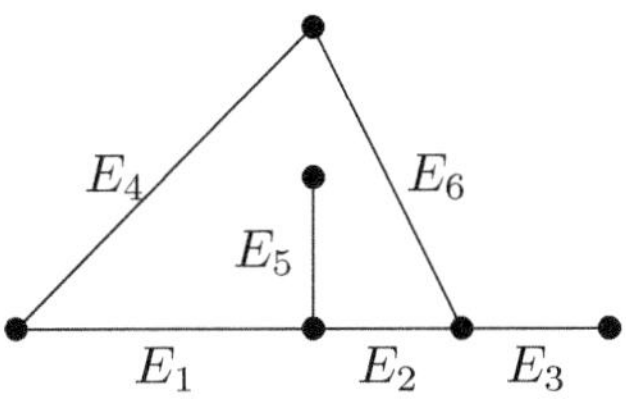

FIGURE 7.19 Planar graph with edges labeled

Theorem 7.8. *Given a connected planar graph with e edges and v vertices. If $v \geq 3$, then $e \leq 3v - 6$.*

Proof. If $v = 3$, then the only connected graphs are a tree with two edges or triangle with three edges, so the theorem is true in this case. We need to prove the theorem for $v > 3$.

Draw the graph in the plane. Let $F_1, F_2, \ldots, F_f$ denote the faces. We define e_i to be the number of edges on face F_i. Every face has at least three edges. So,

$$3f \leq \sum_{i=1}^{f} e_i.$$

Edges are either on the boundary of two faces or on the boundary of one face, so

$$\sum_{i=1}^{f} e_i \leq 2e.$$

Combining the two inequalities tells us

$$3f \leq 2e.$$

We have a connected planar graph, so $v - e + f = 2$. Using $f = 2 + e - v$, we obtain

$$6e + 3e - 3v \leq 2e.$$

Rearranging gives

$$e \leq 3v - 6.$$

□

We can use this theorem to obtain a result concerning the degrees of vertices that we will need later.

Corollary 7.3. *In any connected planar graph there is a vertex with degree 5 or less.*

Proof. If the graph has less than 6 vertices, then it cannot have a vertex of degree 6 or greater.

Suppose, for a contradiction, that we have a connected planar graph with 6 or more vertices and that every vertex has degree of 6 or greater. The sum of the degrees will be greater than or equal to $6v$. We know by Theorem 7.1 that the sum of the degrees is twice the number of edges. This tells us

$$2e \geq 6v,$$

or, equivalently,

$$e \geq 3v,$$

but this contradicts

$$e \leq 3v - 6.$$

□

7.4 NONPLANAR GRAPHS

By definition, planar graphs can be drawn in the plane in such a way that no edges cross. We can also talk about graphs that can be drawn on the surface of a sphere without edge crossings. These sets of graphs are identical. The fact that planar graphs can be drawn on the sphere without edge crossings is not surprising. Draw the graph on the ground, and then zoom out until you see it as a tiny graph on the surface of the earth. The fact that graphs on the sphere without edge crossings are planar, is more surprising. But it is always possible to use the method we used for converting Platonic solids to planar graphs—delete a face and then stretch and flatten the remaining surface until its part of a plane.

The genus of a closed surface counts the number of 'holes' it has. A sphere has genus 0, a torus has genus 1, a torus with a handle attached has genus 2, and so on. If we are given a graph that is not planar—every drawing of it in the plane has edges that cross. However, it is possible to draw it without edge crossings on a surface of higher genus. If a graph can be drawn without edge crossings on a surface with genus g but not on a surface with lower genus, we say the graph has *genus* g. Planar graphs have genus 0. We will look at some graphs with genus 1.

Graphs with Genus 1

We will be drawing graphs on the torus. To visualize these graphs it helps to picture the torus as a rectangle with opposite sides identified. Figure 7.20 shows a torus along with two circles. The red circle goes around the hole and the blue circle goes through the hole. Any circle drawn in the plane, or on a sphere, has the property that they can be continuously deformed and shrunk down to a point. The two circles on the torus do not have this property. This is an important point, and we will return to it.

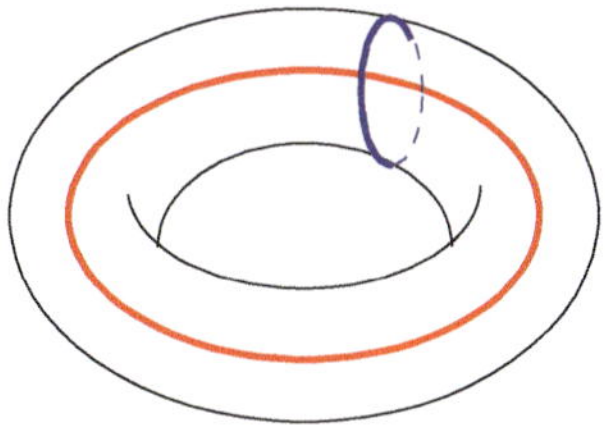

FIGURE 7.20 Torus

If we cut along the blue circle, and then straighten out the tube, we obtain the cylinder shown in figure 7.21. The two blue circles at the end of the tube need to be glued back the obtain the torus. The red line has its ends identified. Remember that it is a circle on the torus.

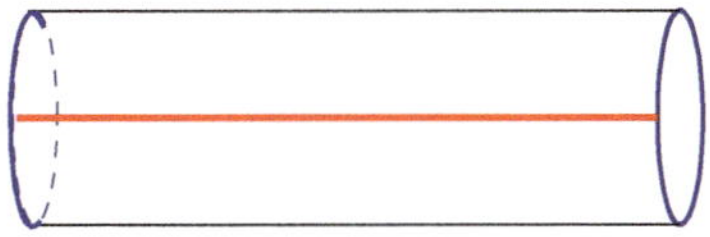

FIGURE 7.21 Torus after cutting along blue circle

We can now cut along the red line and unroll the cylinder to form a rectangle as shown in figure 7.22. The top and bottom edges are both red to help us to remember they, like the blue edges, need to be glued together to obtain the torus.

FIGURE 7.22 Torus after cutting along both circles

We will now try to draw the utilities graph $K_{3,3}$ on the torus. We begin by taking the partial graph depicted in figure 7.13 to obtain figure 7.23. We colored the three regions by yellow, green and white, but we have to be careful. A *face* is something that can be stretched or shrunk, but not cut, to obtain a polygonal face. The yellow and green regions are faces, but the white region is not. The white region corresponds to a torus with a hole cut from it. It contains the red and blue circles that cannot be shrunk down to a point.

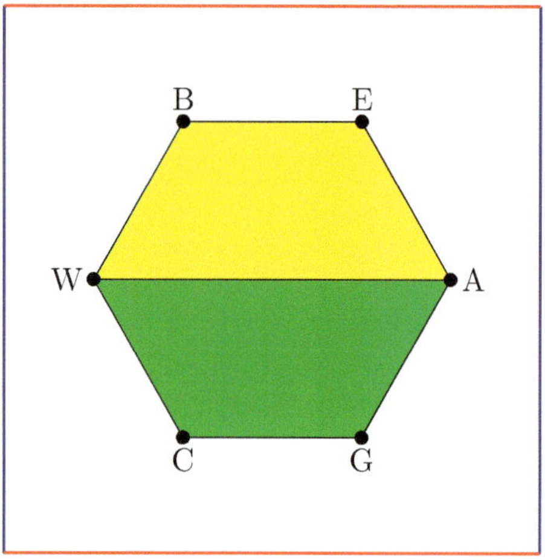

FIGURE 7.23 Part of $K_{3,3}$

We will now connect C to E by going through the hole of the torus. This is shown in figure 7.24. There are three distinct connected regions. Notice that the left of the white region is connected to the right when we glue the two blue sides. However, once again, this white region is not a face. We have chopped the red circle using an edge, but we still have the blue circle.

Finally we add the last edge from B to G. Figure 7.25 shows the graph and the three connected regions. The white region is connected—the top is

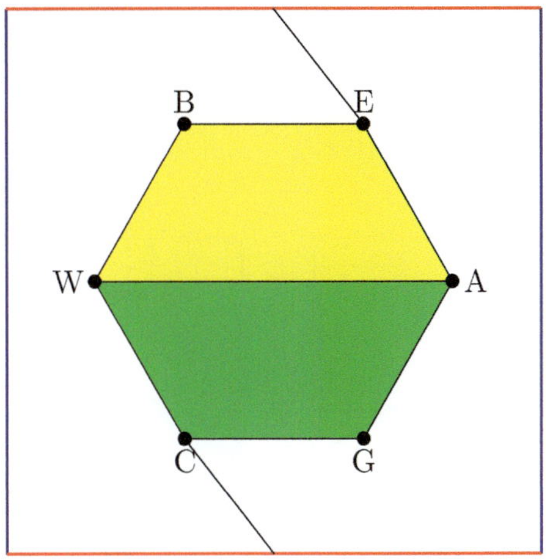

FIGURE 7.24 Adding another edge

glued to the bottom and the left part to the right. However, we have edges that chop through both the red and blue circles. The white region is now a face.

We have constructed a graph of $K_{3,3}$ on the torus. We know $K_{3,3}$ is non-planar, so it has genus 1. The graph has 6 vertices, 9 edges and 3 faces. The Euler characteristic $v - e + f$ is $6 - 9 + 3 = 0$.

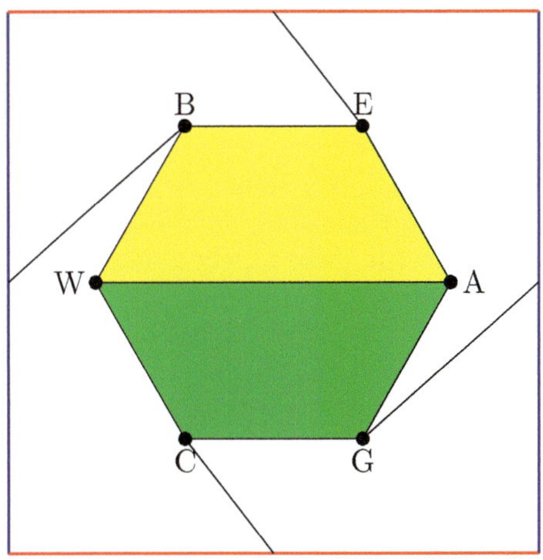

FIGURE 7.25 The graph of $K_{3,3}$

Whenever we draw a graph of a surface without edge crossings it will always chop the surface into regions. If the graph has a genus that is less than the surface, then at least one of the regions contains a 'hole' and, consequently, is not a face. However, the graph and surface have the same genus, the graph can be drawn on the surface in such a way that all the regions will be faces.

Figure 7.26 shows K_5 drawn on the torus without edge crossings. Like $K_{3,3}$, it is genus 1. It has 5 vertices, 10 edges and 5 faces, and Euler characteristic 0.

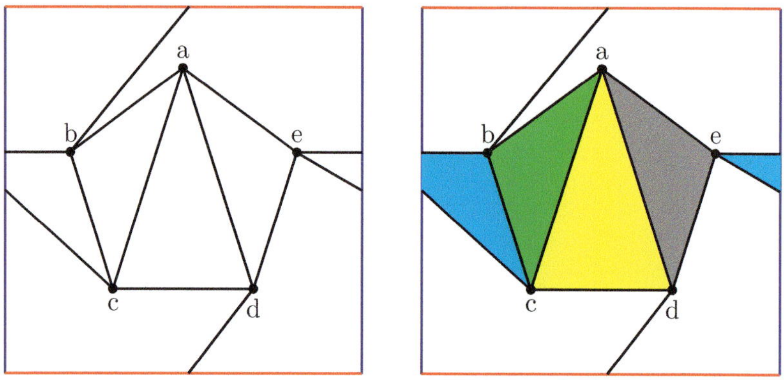

FIGURE 7.26 Graph of K_5 without and with faces colored

Every planar graph has Euler characteristic 2. Every graph of genus 1 has Euler characteristic 0. Theorem 7.7 generalizes the result for graphs of any genus.

Theorem 7.9. *Any connected graph of genus g has Euler characteristic* $2-2g$.

From this theorem we can get a formula for the genus of a surface that involves the number of vertices, edges and faces of a graph. In *algebraic topology* this fact is used when shifting the focus to surfaces. Given a surface, we can triangulate it, then calculate the genus from the triangulation. This approach generalizes to higher dimensions, For a three-dimensional object, we can chop it into tetrahedrons. In general, these higher dimensional analogs of triangulations are called *simplicial complexes*.

7.5 COLORING

England is divided into regions called counties. Maps of England showing the counties usually have a coloring scheme in which neighboring counties have different colors. Francis Guthrie wondered what was the minimum number of colors needed. He looked at various maps and found examples of where four colors were needed, but none that needed five or more colors. He conjectured that any map could be colored with four colors. Guthrie's brother was studying mathematics with Augustus De Morgan, and in 1852 showed him Francis Guthrie's work. De Morgan found the problem interesting. This problem became known as the *four color problem.* Figure 7.27 shows an example of a map with four regions that needs four colors.

FIGURE 7.27 Four regions needing four colors

Given a map, we can draw a graph by shrinking the regions to vertices and drawing an edge between vertices when the corresponding regions share a border. Then color the vertices according to the color of the regions.

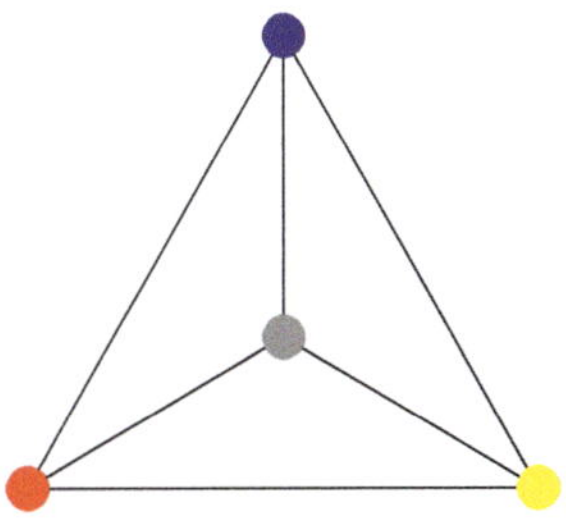

FIGURE 7.28 Graph corresponding to regions in figure 7.27

Chromatic Number

We say two vertices in a graph are *adjacent* if there is an edge that connects them. A *vertex coloring* of a graph assigns colors to each of the vertices in such a way no adjacent vertices have the same color. The *chromatic number* is the smallest number of colors needed for a vertex coloring.

If a graph has at least one edge, then since the two vertices it connects to have different colors, it must have chromatic number of at least 2.

Exercise 7.8.* *Show any tree with at least one edge has chromatic number 2.*

Exercise 7.9. *Show any bipartite graph has chromatic number 2.*

Exercise 7.10.* *Show that the complete graph on v vertices has chromatic number v.*

The four color problem can be restated as saying that every planar graph has chromatic number of 4 or less. In 1879, Alfred Kempe published a paper in which he claimed to have proved this. In 1890, Percy John Heawood showed the purported proof was incorrect. Heawood, using the ideas from Kempe, was able to prove the five color theorem.

The five color theorem

We will follow the work of Kempe and Heawood to prove the five color theorem.

Theorem 7.10. *The chromatic number of any planar graph is 5 or less.*

Proof. It is possible that the planar graph is disconnected. If it is, its chromatic number will be the largest chromatic number of its connected subgraphs, so we only need to prove the result for connected graphs.

We will use a proof by induction on the number of vertices. Clearly, the chromatic number of any planar graph with 5 or fewer vertices cannot be greater than 5. We will now show that if a planar graph with n vertices can be colored with 5 or fewer colors, then so can any planar graph with $n+1$ vertices.

Suppose we are given a planar graph G with $n+1$ vertices. Corollary 7.3 tells us there must be a vertex V of degree five or less. Form the graph G' by removing V and any edges incident to it. Since G' has n vertices it can, by the inductive hypothesis, be colored with five or fewer colors. Choose a coloring for G' that uses at most five colors.

We will now add the vertex V and its edges back to obtain G. If the degree of V is less than 5, then V will be adjacent to at most 4 vertices. These vertices will have at most 4 colors, so there is at least one color left to color V.

We now have to consider the case when the degree of V is 5. The vertex V is adjacent to 5 vertices. If two or more of the adjacent vertices have the same color, there will be a color left over for V.

We now need to see what to do in the case when the 5 adjacent vertices are colored with all 5 colors. We will label the vertices in counterclockwise order as V_1, V_2, V_3, V_4, and V_5 and let the corresponding colors be black, white, red, green, yellow. Figure 7.29 depicts the vertices and colors. (The vertex V is shown as gray. This is not another color, but just meant to indicate the absence of a color.)

The colors associated with vertices V_1 and V_3 are black and red, respectively. We are now going to construct a connected subgraph consisting of only black and red vertices. We begin at V_1 and add any adjacent vertices that are colored red or black along with their associated edges. We keep on adding any black or red vertices adjacent to any vertices we have already added. We also add the corresponding edges. This process eventually results in a connected subgraph $H_{1,3}$ of G containing V_1 with all vertices colored red or black. This subgraph is maximal in the sense that if there is a red or black vertex that is not in $H_{1,3}$ it is not adjacent to any of the vertices in $H_{1,3}$. We can then flip the coloring of the vertices in H and obtain a new vertex coloring of G. After we flip the colors, V_1 will be colored red.

There are now two possibilities. If V_3 is not in $H_{1,3}$, it will remain colored red, and this frees up black for V, and we have a coloring for G. If V_3 is in H, then its color is flipped to black, and V still has adjacent vertices using all five colors. This second possibility seems problematic, but if it occurs we can deduce there is a path from V_1 to V_3 that only has red and black vertices.

If flipping the vertex coloring results in both the colors of V_1 and V_3 being flipped, we can perform the same trick using vertices V_2 and V_4. This flips the color of V_2 from white to green. If the color of V_4 is still green, it frees up white to color V.

At this stage there seems to be a problem. What happens if when we flip the white-green colors V_4 gets flipped to white? If V_4 gets flipped, there will be a path from V_2 to V_4 with only green and white vertices. We will show this cannot occur.

Since our graph is planar, there cannot be a path from V_1 to V_3 and another non-intersecting path from V_2 to V_4. If a first attempt at flipping the vertex coloring results in both the colors of V_1 and V_3 being flipped, the second attempt will change the color of V_2 to green , but the color of V_4 does not get flipped. It remains green. We can then color V white to obtain a coloring of G.

□

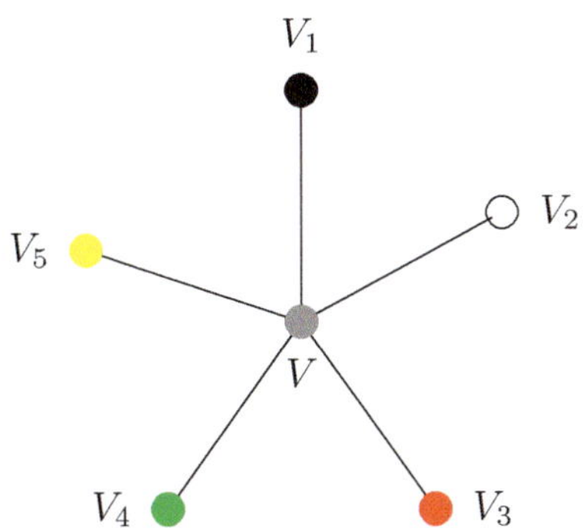

FIGURE 7.29 Vertex V with adjacent vertices

The four color theorem

The four color theorem was finally proved in 1976 by Kenneth Appel and Wolfgang Haken. This proof was somewhat controversial. Appel and Haken reduced the problem to checking whether 1476 graphs could colored using four colors. They then used a computer program to find colorings of these graphs. It took over 1,000 hours of computer time.

In 1996, Neil Robertson, Daniel Sanders, Paul Seymour, and Robin Thomas simplified the proof. They reduced the number of graphs that needed checking to 633. They also used a computer but with a more efficient algorithm.

The theorem has been proved, but there is no known proof that does not use a computer.

7.6 EULER CIRCUITS AND HAMILTONIAN CYCLES

There are various ways of traversing a graph. We always start at a vertex then go along an edge connected to the vertex to the adjacent vertex, and repeat the process, ending at another vertex. This gives a sequence of vertices and edges. We use the term *path* if we are not allowed to repeat any edge or any vertex. Sometimes we want to be less restrictive. If we are allowed to repeat vertices, but not edges, we call the sequence a *trail.* A trail that begins and ends at the same vertex is called a *circuit.* An *Eulerian trail* is a trail in a graph that uses all the edges, and an *Eulerian circuit* is a circuit that uses all the edges.

It is straightforward to tell whether or not a graph has an Eulerian trail or circuit. We begin with circuits.

Theorem 7.11. *A connected graph has an Eulerian circuit if and only if every vertex has an even degree.*

Proof. There are two parts to the proof. We start by showing that if the graph has an Eulerian circuit, then every vertex has even degree. After we will show the converse: if every vertex has even degree, then we can find an Eulerian circuit.

Suppose we are given a graph and an Eulerian circuit. We can start at any vertex V and follow the trail. Since we have a circuit, the trail will also end at V. The trail uses all the edges, so it must contain all the vertices. Whenever the trail enters a vertex W along an edge, it must leave along another edge, so W must have even degree. The only exception is V. We are thinking of the trail ending at V, but the trail also begins at V, so it must also have even degree.

For the converse, suppose we are given a connected graph and every vertex has even degree. We pick a vertex V. We construct a trail by choosing an edge we have not previously included. Eventually the process will end. It cannot end at any other vertex but V, because they all the vertices have even degree. So, we have a circuit, but it might not contain every edge. We will denote this circuit by C_1. If not every edge in the graph is in the circuit, then, because our graph is connected, there must be an edge not C_1 that is incident to a vertex in our circuit. Choose one of these vertices, W. Starting at W, construct a trail by using edges that are not in C_1. As before, the process will end at the starting vertex W.

We construct the circuit C_2 by starting at V following along C_1 until we get to W. Then we go around the circuit we just found, returning to W. After this we continue along C_1 until we get back to V.

If C_2 contains all the edges, we are done. If not, there will be an edge that we have not used incident to a vertex in C_2. We repeat the process of finding a circuit consisting of edges we haven't previously used and adjoining it.

We keep repeating the process until all the edges are used and we have an Eulerian circuit.

□

An *open* trail is a trail where the beginning and ending vertices are distinct.

Theorem 7.12. *A connected graph has an open Eulerian trail if and only if it has exactly two odd vertices.*

Exercise 7.11. *Prove Theorem 7.12 by adding an additional vertex connected by two edges to the vertices of odd degree.*

An Eulerian circuit is a trail through the graph which uses each edge exactly once. We can also ask about visiting vertices exactly once. A *Hamiltonian path* is a path through a graph that contains all of the vertices. If the starting vertex is the same as the ending vertex, we have a *Hamiltonian cycle.*

These are named after William Rowan Hamilton. Hamilton did important work in both mathematics and physics. He also invented the *icosian game* in 1857. This was a planar graph representing an icosahedron. The vertices were labeled with the names of cities. The challenge was to travel along the edges visiting each city once and only once, before ending at the starting city. Hamilton was interested in symmetries of the icosahedron, but the game was a commercial venture. Hamilton sold the rights for 25 pounds to Jaques of London, who then marketed throughout Europe. It was an easy puzzle to solve, and not a commercial success.

We know that if a connected graph has an Eulerian circuit if and only if every vertex has even degree. To prove this, we used an algorithm for finding the circuit. We might expect there is a simple criterion for telling whether a connected graph has a Hamiltonian cycle, and a simple algorithm for finding one. Neither of these are true. There is no known simple criterion that tells us whether or not a graph has a Hamiltonian cycle. There are algorithms for finding these cycles, using extensive searches through all the possibilities, but they are not fast.

A related question is the *traveling salesperson* problem. In this, a graph is given with the vertices representing cities. The edges between the vertices represent roads. Each edge has a number giving the distance between the cities. The problem is to find the shortest route that begins and starts at the same city and visits every other city exactly once. This problem has many practical applications and has been extensively studied. An important outstanding question is whether there is an algorithm for solving the Hamiltonian path problem, or traveling salesperson problem, in polynomial time. We will come back to this in the last chapter.

Multigraphs

A multigraph is a graph except we allow the possibility of having more than one edge between two vertices. The graph for the Königsberg bridge problem, figure 7.2, is an example. We extend our definitions for trails from graphs to multigraphs in the obvious way: An Eulerian trail is a trail through the multigraph that uses each edge once; an Eulerian circuit is an Eulerian trail that starts and ends at the same vertex.

The criteria for Eulerian circuits and open trails can be generalized from graphs to multigraphs. We leave it to the reader to check this.

Theorem 7.13. *A connected multigraph has an Eulerian circuit if and only if every vertex has an even degree.*

Theorem 7.14. *A connected multigraph has an open Eulerian trail if and only if it has exactly two odd vertices.*

The graph for the Königsberg bridge problem has four vertices. All of them have odd degree, so it does not have an Eulerian circuit or an open Eulerian trail.

7.7 GAUSSIAN CURVATURE AND THE GAUSS–BONNET THEOREM

The Gauss–Bonnet theorem properly belongs to the area of mathematics known as *Differential Geometry*, where calculus is used to frame and solve geometrical problems. The reason for including it in a chapter on Graph Theory is that, as we will shortly see, the theorem also involves the Euler characteristic introduced in Section 7.4.

We start by considering the following question: *how curved is a (planar) curve?* What we are asking for here is a number to assign each point on the curve which indicates the degree of curvature at that point.

A natural approach to the problem is as follows: proceed along the curve at a steady rate and keep track of the angle θ between the tangent line to the curve and the horizontal (see figures 7.30 and 7.31). If θ is changing fast then we are at a place of high curvature (we need a hairpin bend sign!). But there is a natural measure of the rate of change of a function: the *derivative*. We thus define the *curvature* κ at point P on the curve to be

$$\kappa = |\theta'(s)|,$$

where absolute value is included to make curvature a non-negative quantity.

For example, in figure 7.30, $\theta = \theta_0$ is constant, hence $\kappa = 0$. By contrast, for the arc of the circle with radius R depicted in figure 7.31,

$$\theta = \frac{\pi}{2} - \frac{\theta}{R}.$$

Hence $\kappa = 1/R$.

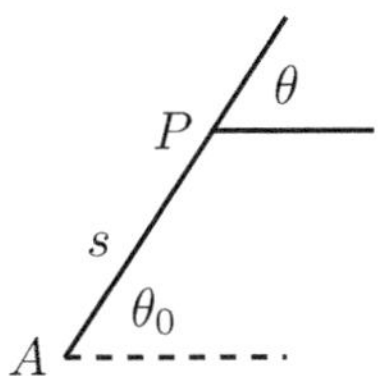

FIGURE 7.30 Tangent on a line

We now turn to the notion of the curvature of a surface. Gauss devised a splendid way to measure this: At any point P on the surface, consider the normal line L to the surface at P (the line perpendicular to the tangent plane). Then consider a plane containing L and rotate the plane about L. As the plane rotates through 180° it cuts the surface in a family of curves. Let

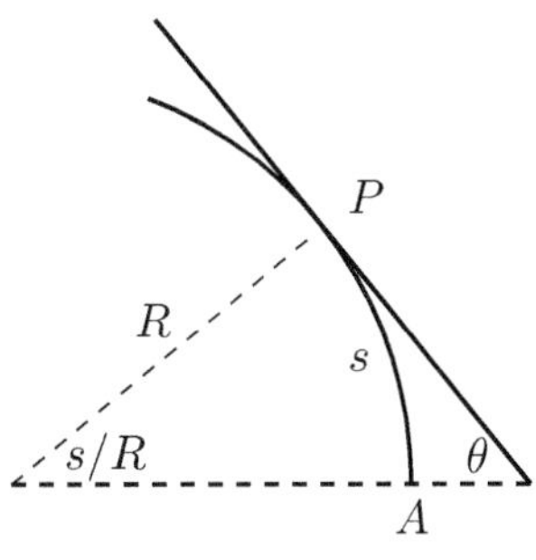

FIGURE 7.31 Tangent on a circle

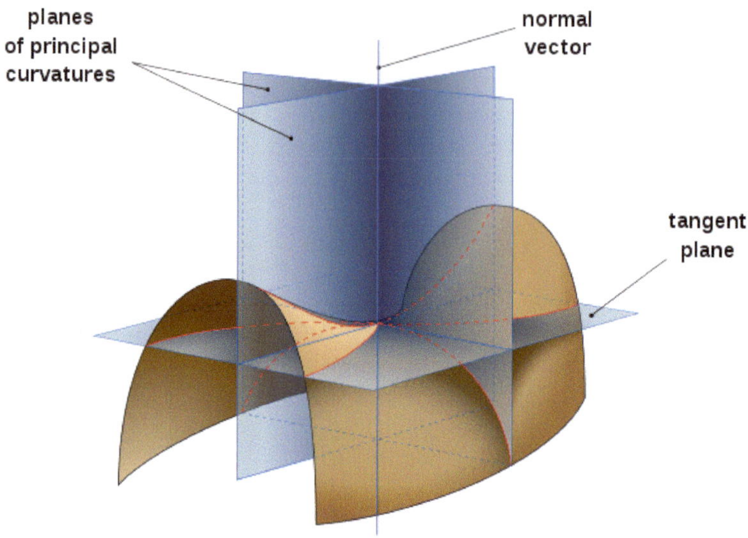

FIGURE 7.32 A saddle with principal curves indicated (author Eric Gaba)

κ_1 and κ_2 denote the smallest and the largest of the curvatures, called the *principal curvatures* (see figure 7.32). The *Gaussian curvature* κ of the surface at P is defined by

$$\kappa = \pm\kappa_1\kappa_2$$

with the convention that the sign is chosen according to whether the principal curves lie on the (positive) same, or opposite (negative) sides, of the tangent plane at P.

A couple of examples. Consider (a region of) the sphere of radius R. Clearly, every normal plane cuts the sphere in a great circle, of radius R. Thus $\kappa_1 = \kappa_2 = 1/R$, hence $\kappa = 1/R^2$.

Now consider a piece of a cylinder of radius R. Then, at any point, the curvature κ_1 in the direction along the cylinder is zero, while κ_2, the curvature transverse to the cylinder is $1/R$; thus $\kappa = 0$. The sphere and the cylinder are examples of surfaces of constant curvature, but in general Gaussian curvature will vary from point to point on a surface, consider, e.g., a squashed balloon.

The most fundamental property of Gaussian curvature is that it is *intrinsic* to the surface. By this is meant the following: imagine wadding up a flat sheet of paper ready to toss into the trash. The sheet prior to wadding, and after, looks quite different. But this difference is, in one sense, illusory, due to the fact that we are looking at a two-dimensional world through 3-dimension eyes. A sheet of paper, although flexible, is *non-elastic*; while it can be bent and twisted, it cannot be stretched or compressed. As a consequence, the distance between any two points on the surface (as measured by the length of the shortest path in the surface between the two points) is unchanged in the wadding.

To a hypothetical two-dimensional resident of the surface, who has no knowledge of three-dimensional space, the flat and the wadded ball of paper would appear identical! We say that the two surfaces are *isometric*. Gaussian curvature is an *isometric invariant*: two isometric surfaces have identical Gaussian curvatures at corresponding points. Thus the wadded up paper has Gaussian curvature zero at every point (the same as the flat plane). On the other hand, the fact that the sphere and the plane have different Gaussian curvatures implies that they are not isometric. This explains why it is impossible to make a flat map of the world that does not distort distances.

The intrinsic nature of Gaussian curvature seems remarkable in light of the definition given above, since the two principal curvatures are generally non-intrinsic. Consider a piece of a cylinder and the region of the plane formed by unrolling so that it is flat. The two surfaces are isometric, yet have different principal curvatures (at least in the direction transverse to the cylinder). The point is that the *product* of the curvatures–0–are in agreement. Gauss proved the isometric invariance of his curvature. He must have been (justly) proud of this work for, of all his mathematical discoveries, he named it his *Theorema Egregium* ("remarkable theorem").

The Gauss–Bonnet theorem relates the Gaussian curvature of a surface and the Euler characteristic of the surface.

Theorem 7.15. *(Gauss–Bonnet) Let S denote a closed compact orientable surface with Gaussian curvature κ and Euler characteristic χ.*[1] *Then*

$$\int_S \kappa \, dA = 2\pi\chi. \tag{7.1}$$

The integral in (7.1) measures the "total curvature" on the surface. Consider the case when S is a sphere with radius R. Since $\kappa = 1/R^2$ is constant, the integral is

$$\frac{1}{R^2} \times A,$$

where A is the surface area of the sphere. Having no holes, the sphere has Euler characteristic $\chi = 2$. The Gauss–Bonnet theorem thus yields

$$A = 4\pi R^2$$

the well-known formula for the surface area of a sphere with radius R.

[1] To say that a surface is closed and compact means that the surface is finite in extent and without boundary, such as the sphere or the torus, as opposed to a bottle, which has a boundary at the rim. "Orientable" is a technical condition needed to make sense of the area element dA in the formula. An orientable surface is one with a continuously defined normal direction. It amounts to saying that the surface has an inside and an outside. For example, the sphere and the torus are orientable surfaces, while the Möbius strip is not.

Recall that $\chi = 2(1 - g)$ where g is the genus of the surface, i.e., the number of holes.

The Gauss–Bonnet theorem is one of the most amazing results in mathematics. To get a feel for just how amazing, consider a spherical balloon. It has total curvature 4π. Now suppose the balloon is twisted into the shape of a balloon dog. The curvatures obviously change in the transformation from balloon to dog, probably at every single point. But somehow, it all evens out in the long run and the total amount of curvature on the dog again ends up being 4π!

Is this plausible? We offer a few examples in this direction. Consider first what happens when we blow air into a balloon. The curvature of the balloon decreases at every point as the balloon inflates, at the same time the surface of the balloon increases. These opposing tendencies counteract each other and the total curvature on the balloon remains the same, i.e., 4π.

Consider a donut (torus), which has genus 1 and hence Euler characteristic 0. Gauss–Bonnet tells us that the total curvature is zero. Why does the sphere have positive curvature, while for the torus, the curvature is zero? Because the torus, unlike the sphere, has both regions of positive and negative curvature. The regions of negative curvature occur around the hole, as illustrated in figure 7.33. The positive and negative curvatures cancel each other out, and the total curvature is zero.

Suppose we punch another hole in the torus, to create a surface with genus 2 (figure 7.34). The presence of the extra hole creates more negative curvature on the surface, and Gauss–Bonnet tells us that the total curvature is now -4π. And so it goes ...

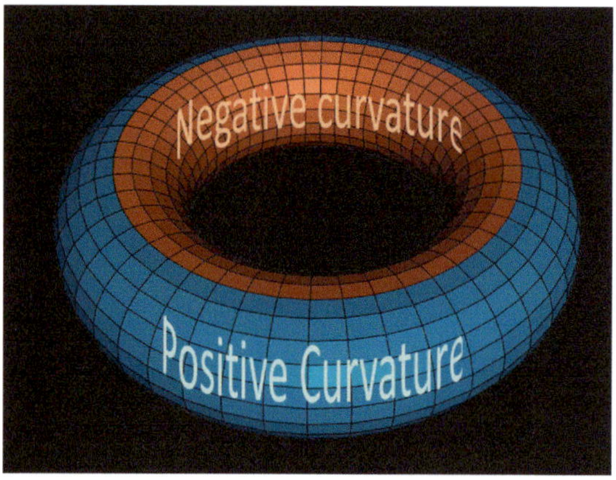

FIGURE 7.33 Negative and positive curvature on a torus (author Gregors)

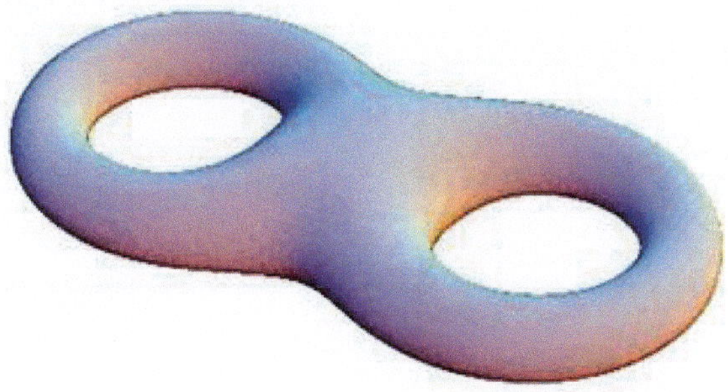

FIGURE 7.34 A two-holed torus (from Wolfram MathWorld)

Suggestion for Further Reading

Nora Hartsfield, Gerhard Ringel *Pearls in Graph Theory.* Dover Publications, 2013

Richard J. Trudeau *Introduction to Graph Theory.* 2nd Edition. Dover Publications, 1994.

CHAPTER 8

Probability

The subject of Probability, with its origins in games of chance, is almost as old as humanity itself. We begin this chapter with a historical introduction to the subject.

Sections 8.2–8.4 deal with discrete probability, based on counting arguments. The techniques are used to solve several problems of general interest, e.g., calculating odds in poker, analyzing a two party gambling game, and the so-called problem of points (how to divide the stakes when a gambling game is interrupted before its conclusion). Also discussed is the random walk in one and two-dimensions.

In the second half of the chapter, the reader is introduced to continuous probability models. Topics covered include the normal distribution, the strong law of large numbers and the central limit theorem. The role of probabilistic methods in mathematical statistics is indicated by means of an example. The chapter concludes with an application of probability to number theory in the form of the normal number theorem.

8.1 HISTORY

Almost every primitive culture engaged in some form of dice play. The predecessor of the die was a tarsal bone (the *astragalus*) taken from the hind foot of a hoofed animal. The astragalus can rest on only four sides, two broad and two narrow, differing sufficiently in appearance to make them distinguishable.

Dice games are thought to have evolved from divination rites. A question was posed to a god by a priest; the dice were cast on the sacred ground, and the god's answer deduced from the outcome. One of the games in ancient Greece consisted of the simultaneous casting of four astragali and noting which sides fell uppermost. The throw of highest value was generally considered to be the one that showed the four different faces of the four bones (the "throw of Venus").

DOI: 10.1201/9781003470915-8

The six-sided die may have been obtained by grinding down the astragalus until a cube was formed. The earliest known dice were discovered in northern Iraq, and date from 3000 BCE. They were made from fried pottery, and the faces marked with from one to six dots in the fashion of today. Gambling with dice was a commonplace recreation in ancient Rome. The Emperor Claudius was supposed to be so devoted to dicing that he wrote a book on the subject, and used to play while riding his chariot. (A precursor to our present day practice of texting while driving?) Gambling was declared to be immoral by many civilizations and both the Jews and the Christians had laws forbidding it.

The origin of card play is variously attributed to the Egyptians, the Chinese, and the Indians. The first playing cards made in Europe were very fancy hand-painted affairs, available only to the upper classes. With the advent of printing, playing cards became widely available. Around 1500, the French developed the present suits of spades, hearts, clubs, and diamonds.

Cardano's *Liber de Ludo Aleae*, published in 1663, is generally considered to be first work linking games of chance to mathematics. Herein, Cardano gave the definition of the probability of an event as the number of ways of achieving the event, divided by the totality of possible outcomes. Cardano investigated the probabilities of casting astragali and one or several dice, and he calculated the probabilities of the occurrence of certain card combinations in the game of *pimero* (similar to poker). A passage in a chapter entitled "On the Cast of One Die" reads

I am as able to throw 1, 3, or 5, as 2, 4, or 6. The wagers are therefore laid in accordance with this equality if the die is honest and, if not, they are made so much the larger or smaller in proportion with the departure from true equality.

In 1494, Fra Luca Pacioli posed the following problem:

A team plays ball so that a total of 60 points is required to win the game and the stakes are 22 ducats. By some incident, they cannot finish the game and one side has 50 points, and the other 30. What share of the prize belongs to each side?

Fra Luca's solution to this so called "problem of points" was that the stakes should be divided in the proportion 5:3, the ratio of points already scored. Cardano pointed out that this reasoning is incorrect since it fails to take into account the number of points yet to be scored. He then proceeded to give his own solution, also incorrect. Perhaps the most remarkable feature of Cardano's solution was the acerbic manner in which it was presented, "And there is an evident error in the determination of the game problem as even a child should recognize, while he [Fra Luca] criticizes others, and praises his own excellent opinion." Tartaglia (Cardano's bitter enemy, it will be recalled in the matter of publication of the solution to the cubic equation) then weighed in with his own solution, also wrong. The problem of points was proving to be a rather hard nut to crack. The correct solution, that the stakes should be awarded in the ratio 7:1, was only discovered by Blaise Pascal (1622–1663) in 1654. (It will be presented in the next section.) Fermat also discovered the solution around the same time. In the course of their ensuing correspondence, Pascal and Fermat laid the foundations for a mathematical theory of probability.

James Bernoulli (1654–1705) made a major contribution to probability theory in his *Ars Conjectandi*, published in 1713. In this work Bernoulli proved a version of the Law of Large Numbers: suppose E is an event that occurs with probability p. Then the proportion of such events that occur in a sequence of independent trials approaches p as the number of trials becomes arbitrarily large. This is crucial for the application of probability in statistics. The relevance of the law is that it links the theoretical probability p of the event with the proportion of times the event is likely to occur in actual practice.

Another major figure in the history of probability theory is Pierre Simon Laplace (1749–1827). Laplace wrote several influential books on the subject, including his *Thèorie Analytique des Probabilités*, published in 1812. Herein, Laplace gave a formal definition of classical probability along the lines suggested by Cardano and formulated the fundamental principles of the subject. Further important contributions followed, by Daniel Bernoulli, Poisson, Chebyshev, and Markov.

FIGURE 8.1 A. N. Kolmogorov, courtesy of Konrad Jacobs, Erlangen

Perhaps the most important probabilist of the twentieth century was the Russian mathematician Andrey Nikolaevich Kolmogorov (1903–1987). In his fundamental work *Foundations of Probability*, Kolmogorov developed a series of axioms for the subject and set the stage for a myriad of developments that were to follow. Among Kolmogorov's early contributions to probability theory were new proofs of the two fundamental limit laws, the Strong Law of Large Numbers and the Central Limit Theorem. Another highly significant achievement of Kolmogorov was an equation establishing a link between probability theory and Fourier's theory of heat.

Further impetus to the development of modern probability theory was the desire to provide a mathematical foundation to the random movement of particles observed by Robert Brown in 1827. This was the subject of one of Albert Einstein's famous papers of 1905, credited with convincing the scientific world of the existence of molecules. A rigorous model of Brownian motion (the Wiener process) was constructed by Norbert Wiener in 1923. Wiener's work was to have profound repercussions for mathematics and science. A striking example is the theory of stochastic integration based on the Wiener process, created in the 1940s by the Japanese mathematician Kiyosi Itô.

8.2 DISCRETE PROBABILITY

The classical definition of probability originated by Cardano, is as follows. Suppose there are N possible outcomes of a certain action, all equally likely to occur, and a certain event E consists of n of these outcomes. Then the probability of E, denoted by $P(E)$ is given by

$$P(E) = \frac{n}{N}.$$

Suppose, for example, we toss a fair die. There are six equally likely outcomes, 1,2,3,4,5,6. The event E, that the dice throw results in an even number, contains three possible outcomes: 2, 4, 6. Hence

$$P(E) = \frac{3}{6} = \frac{1}{2}.$$

In general, outcomes in an experiment do not necessarily need to have equal probabilities (e.g., a biased coin or a weighted die). In this case, the probability of an event must be calculated by adding together the probabilities of the separate outcomes that comprise the event.

Based on these notions, Laplace worked out the basic laws of probability. These laws, and further theorems in probability, are most easily framed in the language of sets. The S set of all possible outcomes of an experiment is referred to as the *sample space* of the experiment. *Events* are subsets of S. For two events A and B, the event that *either* A or B occurs (where the scenario allows that both A and B occur) is the set the set $A \cup B$, while the event that *both* A and B occur is $A \cap B$.

Laplace's laws of probability are the following.

1. Let A and B denote *mutually exclusive* events, i.e., $A \cap B = \phi$, the empty set. Then

$$P(A \cup B) = P(A) + P(B). \tag{8.1}$$

2. Suppose A and B are *independent* events, i.e., the occurrence of one event has no bearing on the probability of the other event occurring. Then

$$P(A \cap B) = P(A) \cdot P(B). \tag{8.2}$$

Exercise 8.1. *Use (8.1) to prove that for any two events A and B*

$$P(A \cup B) + P(A) + P(B) - P(A \cap B). \tag{8.3}$$

Hint: decompose the set $A \cup B$ into disjoint sets.

To take an example, consider the experiment of tossing a fair die and a fair coin simultaneously. The sample space S consists of 12 elements, all equally likely. That is,

$$\begin{aligned} S = \{&(1,H),(2,H),(3,H),(4,H),(5,H),(6,H),(1,T),(2,T),\\ &(3,T),(4,T),(5,T),(6,T)\}. \end{aligned}$$

Let A be the event that the die results in an even number, B that the coin comes up Heads, and C that the die results in either 1 or 5. Note that A and C are mutually exclusive and A and B are independent. We have

$$\begin{aligned} &P(A \cup C)\\ &= P(\{(2,H),(4,H),(6,H),(2,T),(4,T),(6,T),(1,H),(5,H),(1,T),(5,T)\})\\ &= 10/12 = 3/6 + 2/6 = P(A) + P(C),\\ &P(A \cap B) = P(\{(2,H),(4,H),(6,H)\}) = 3/12 = 3/6 \cdot 1/2 = P(A) \cdot P(B). \end{aligned}$$

Permutations and Combinations

The above calculations of probabilities relied upon an explicit listing of the sample space and of the events in question. This is unfeasible when the sample space is very large, and unnecessary if the outcomes have equal probabilities. Then we do not need to know which particular outcomes comprise a given event, just the *number* of outcomes. We introduce some methods to determine this.

We begin with the following elementary observation, which goes by the name of the *multiplication principle.* Consider an experiment consisting of k steps, in which there are n_1 ways to perform step 1, n_2 ways to perform step 2, ..., n_k ways to perform step k. Then there are $n_1 n_2 \dots n_k$ ways to perform the entire experiment.

Suppose, e.g., that when dressing in the morning, you can choose between 5 tops, 3 pairs of pants, and 2 pairs of shoes. This makes for $5 \times 3 \times 2 = 30$ possible outfits in all.

A *permutation* is a selection of items where the order of selection matters, i.e., selecting the same objects, but in a different order, constitutes a different selection. As an example, suppose it is required to choose three soccer players from a group of 7 players to serve as a defender, a midfield player, and a forward. There are 7 options for the defender; this leaves 6 options for the midfield player, and after these have been chosen, 5 options for the forward. According to the multiplication principle, there are $7 \times 6 \times 5 = 210$ possible selections.

In similar fashion, the number of permutations of k objects chosen from a group of n objects is

$$\begin{aligned} &n \times (n-1) \times (n-2) \times \cdots \times (n-k+1) \\ &= \frac{n(n-1)(n-2)\ldots(n-k+1) \times (n-k)(n-k-1)\ldots 1}{(n-k)(n-k-1)\ldots 1} = \frac{n!}{(n-k)!}. \end{aligned}$$

This quantity is denoted ${}_nP_k$.

By contrast, a *combination* is a selection of objects where the order of the selection is not an issue. Consider, for example, the number of ways of choosing 3 letters from the set $\{a, b, c, d, e\}$. There are $5 \times 4 \times 3 = 60$ permutations of the letters. Let us arrange the permutations into groups, where each group contains the same 3 letters. We get an array as follows.

$$\begin{array}{cccc} abc & abd & \ldots & cde \\ acb & adb & \ldots & ced \\ bac & bad & \ldots & dce \\ bca & bda & \ldots & dec \\ cab & dab & \ldots & ecd \\ cba & dba & \ldots & edc. \end{array}$$

Note that each column corresponds to a single combination. In each column there are $3! = 6$ entries (the number of permutations of 3 letters). Thus there are $60/6 = 10$ columns, and hence combinations.

In general, there are

$${}_nP_k/k! = \frac{n!}{k!(n-k)!}$$

combinations of k objects chosen from n, The number of combinations is denoted ${}_nC_k$, or more commonly, $\binom{n}{k}$.

We apply this formula to calculate some odds in the game of poker, where 5 cards are dealt from a deck of 52 cards. There are $\binom{52}{5}$ possible poker hands. A "full house" is a hand consisting of three cards in one denomination and two cards in another (for example 5, 5, 5, 8, 8). How many full houses are there? Firstly, there are ${}_{13}P_2 = 13 \times 12$ ways to choose the denominations. Once this has been decided, there are then $\binom{4}{3}$ and $\binom{4}{2}$ ways to choose the cards in these denominations, resulting in ${}_{13}P_2\binom{4}{3}\binom{4}{2}$ possible full houses. Hence the probability of a full house is

$$\frac{{}_{13}P_2\binom{4}{3}\binom{4}{2}}{\binom{52}{5}} = \frac{3744}{2598960} = 00144.$$

Note that in choosing the denominations, order matters (3 Jacks and 2 Sevens is a different hand to 2 Jacks and 3 Sevens).

Let's calculate the probability of "two pairs" (two cards in one denomination, two cards in a different denomination, and an odd card). The number of possible choices of denominations for the pairs is $\binom{13}{2}$. (This time it's a *combination* because of the symmetry of the situation.) There are then $\binom{4}{2}$ and $\binom{4}{2}$ ways to be dealt the cards in these denominations, and 44 choices for the odd card. We conclude that the probability of the hand is

$$\frac{44\binom{13}{2}\binom{4}{2}\binom{4}{2}}{\binom{52}{5}} = .04225.$$

Exercise 8.2[⋆] *Find the probability of a straight in poker (5 consecutive cards).*

We can use the formula for combinations to prove the binomial theorem: consider foiling the binomial

$$(a+b)^n = \underbrace{(a+b)(a+b)\dots(a+b)}_{n}$$

The resulting terms have the form $a^k b^{n-k}$, $k = 0, 1, \dots, n$. This particular combination of powers of a and b arises when we multiply a in k of the brackets by b in the remaining $n-k$ brackets. There are $\binom{n}{k}$ ways to choose the brackets, hence $\binom{n}{k}$ terms of this form. Thus

$$(a+b)^n = \sum_{k=0}^{n} \binom{n}{k} a^k b^{n-k}.$$

Bernoulli, Binomial, and Geometric Random Variables

We introduce some terminology. A *random variable* X is a quantity which takes on a variety of numerical values, depending on chance. The set of probabilities

$$P(X = j)$$

with which X assumes its possible values, is called the *(probability) distribution* of X.

For example, let X be the throw of a fair die. The possible outcomes are $1, 2, 3, 4, 5, 6$, and the distribution of X is

$$P(X = k) = 1/6,\ 1 \leq k \leq 6.$$

Suppose, however, that the die is weighted so that one is twice as likely to throw an even number as to throw an odd number, and the even and odd numbers are equally likely among themselves. The distribution of X is then

$$P(X = k) = \begin{cases} \frac{2}{9}, & k = 2, 4, 6, \\ \\ \frac{1}{9}, & k = 1, 3, 5. \end{cases}$$

Random variables X and Y are said to be *independent* if the events $\{X = i\}$ and $\{Y = j\}$ are independent for all i and j, i.e.,

$$P(X = i \text{ and } Y = j) = P(x = i)P(Y = j).$$

A random variable which assumes only two possible values, is said to be of *Bernoulli* type (It is usual to refer to the outcomes as "success" and "failure.") An example would be a coin flip, and to make it more interesting let's assume it's a weighted coin with P(Head) = .75 and P(Tail) = .25.

Suppose we to flip the same coin 10 times. What is the probability of throwing exactly 6 Heads? The answer is 0.0729. How is this number arrived at? Consider the scenarios which result in 6 Heads and 4 Tails in 10 flips. A typical one is

$$HTTHHHTHTH.$$

Since the coin flips are independent, the probability of this particular string is

$$.75 \times .25 \times .25 \times \cdots \times .75 = (.75)^6(.25)^4.$$

How many such strings exist (with 6 Heads and 4 Tails)? The answer is $\binom{10}{6}$, the number of ways of choosing 6 out of the 10 slots to position the Heads. Therefore the probability of 6 Heads is

$$\binom{10}{6}(.75)^6(.25)^4 = 0.0729.$$

The successive coin flips here are referred to as *Bernoulli trials.*

The number of successes X in a sequence of n independent Bernoulli trials where the probability of success each time is p and (failure $q = 1 - p$) has distribution

$$P(X = k) = \binom{n}{k}p^k q^{n-k},\ k = 0, 1, \ldots, n.$$

X is said to have a *binomial* distribution with parameters n and p.

Exercise 8.3. *The binomial distribution has many practical applications. For example, consider the number of patients X in a group of 20 patients who experience cures in a series of drug treatments, where the probability of success each time is 80%. Find $P(X = 17)$.*

Consider the sequence of the Bernoulli trials, and let X now denote the number of trials required until the first success is obtained. For any $k = 1, 2, 3, \ldots$, there is a single scenario that results in $X = k$, and it is the string

$$\underbrace{FF\ldots F}_{k-1}S.$$

Thus

$$P(X = k) = q^{k-1}p, \quad k = 1, 2, 3, \ldots.$$

Then X is said to be *geometric* with parameter p.

We will now address the "problem of points," mentioned earlier, and give the solution in a slightly more general setting. Suppose as before, player A has accumulated 50 points and player B, 30 points when the game is interrupted. We wish to determine the probability that A reaches 60 points before B and so wins the game. Suppose that on each play, A has probability p of winning the point and player B, probability $q = 1 - p$. We break down event of A winning the game into the series of scenarios whereby B wins k more points before the game ends, with $k = 1, 2, 3, \ldots 29$. This gives rise to a calculation similar to the calculation of the binomial probabilities above. Note, however, that in this instance, the final play is the tenth success for A. The strings that result in this particular scenario are 9 successes and k failures for A in $9 + k$ plays, followed by a final success for A. The probability of any string of this type is $p^{10}q^k$, and there are $\binom{9+k}{9}$ such strings. The probability that B wins k more points before A wins the tenth point is therefore $\binom{9+k}{9}q^k p^{10}$, and the probability that A wins the game is

$$\sum_{k=0}^{29} \binom{9+k}{9} q^k p^{10}.$$

Exercise 8.4.* *Compute this in the classical case $p = q = 1/2$.*

Mean and Variance of a Random Variable

Consider the weighted die where the probability of throwing an even number is 2/9, and an odd number is 1/9. If we throw the die very many times, we might expect that roughly 1/9 of the throws would result in a 1, a 3 and a 5, and 2/9 of the throws would result in a 2, a 4, and a 6. If this actually occurred, then the average of the throws would be

$$(1 \times 1/9) + (2 \times 2/9) + (3 \times 1/9) + (4 \times 2/9) + (5 \times 1/9) + (6 \times 2/9) = 2.333.$$

Motivated by this example, we define, for a random variable X, the *expected value*, (or the *mean*) of X by

$$\sum_k kP(X = k),$$

It denotes the expected "long term average" of X.

The *variance* of a random variable X provides a measure of how spread out the values are likely to be. The variance of X is defined by

$$\sum_k (k - \mu)^2 \cdot P(X = k),$$

where μ is the mean of X.

The mean and variance are denoted by $E[X]$ and $Var(X)$, respectively.

Finally, we define the *standard deviation* of a random variable to be the square root of the variance.

The basic properties of mean and variance are given in the following result.

Theorem 8.1. *For random variables X and Y and constant c*

(i) $E[cX] = cE[X]$
(ii) $E[X + Y] = E[X] + E[Y]$
(iii) $Var(cX) = c^2 Var(X)$
(iv) If X and Y are independent then $Var(X + Y) = Var(X) + Var(Y)$.

Some examples

1. Let X be Bernoulli, with $P(X = 0) = q$ and $P(X = 1) = p$. Then

$$E[X] = 0 \cdot p + 1 \cdot q = p. \quad (8.4)$$
$$Var(X) = (0 - p)^2 \cdot q + (1 - p)^2 \cdot p = p^2 q + q^2 p = pq(q + p) = pq. \quad (8.5)$$

2. Let X be a geometric random variable with parameter p. Taking a small mathematical liberty[1] and using the sum of the geometric series, we have

$$\begin{aligned} E[X] &= p\sum_{j=1}^{\infty} jq^{j-1} = p\sum_{j=1}^{\infty} \frac{d}{dq}q^j \\ &= p\frac{d}{dq}\sum_{j=1}^{\infty} q^j = p\frac{d}{dq}\frac{q}{1-q} \\ &= \frac{p}{(1-q)^2} = \frac{1}{p}. \end{aligned}$$

[1]The liberty is in assuming we can interchange the order of the infinite sum and the derivative. This turns out to be valid, but requires justification.

We can show, by a similar calculation

$$Var(X) = \frac{p}{q^2}.$$

Exercise 8.5.* *Let X be a binomial random variable with parameters n and p. Show that $E[X] = np$ and $Var(X) = npq$. (Hint: write*

$$X = \sum_{i=0}^{n} X_i,$$

where $X_1, X_2, \ldots, X_n$ are independent Bernoulli trials, and use (8.4), (8.5) and Theorem 8.1(ii), (iv)).

In general, it is likely that a random variable will turn up a result within a few standard deviations of its mean. The next result, known as *Chebyshev's inequality*, quantifies this tendency.

Theorem 8.2. *Let X be a random variable with mean μ and standard deviation σ. Then for all $k \geq 1$*

$$P(\mu - k\sigma < X < \mu + k\sigma) \geq 1 - 1/k^2. \tag{8.6}$$

As an example, suppose the scores on a national examination have mean 70% and standard deviation 5%. We want to use Chebyshev's inequality to estimate the proportion of scores 90% or more. Consider the interval from 50 to 90, and note that the gap of 20 on either side of the mean represents 4 standard deviations. The inequality with $k = 4$ implies that the proportion of scores lying in the interval $[50, 90]$ is at least $1 - 1/4^2 = 15/16$. Hence the proportion of scores in excess of 90 can be at most $1/16$.

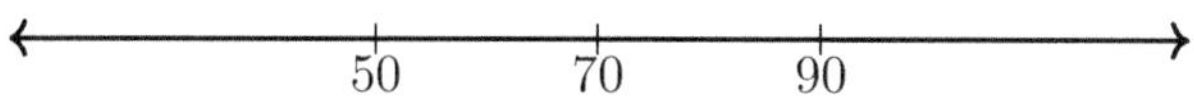

FIGURE 8.2

The following exercise is an example of a statistical test based on the Chebyshev principal. (We will see a much more powerful version of this procedure in Section 8.8.)

Exercise 8.6.* *It is claimed that a certain high school is performing below par on the national exam mentioned above. The average of 100 students from the school on the exam was calculated and found to be 65%.*

First use Theorem 8.1 to show that, if the students' test scores are typical of the national population, then the average

$$\bar{X} = \frac{X_1 + X_2 + \cdots + X_{100}}{100}$$

has mean 70 and standard deviation 0.5. (Consider the 100 test scores as independent random variables with mean 70 and standard deviation 5.) Then apply Theorem 8.2 to show that the claim is quite likely true.

8.3 THE GAMBLER'S RUIN

Consider a gambling game where with each play, the gambler either stands to win a dollar, with probability of p, or lose a dollar, with probability $q = 1 - p$. The gambler decides that they will quit the game if their holdings either reaches a predetermined amount of \$N (in which case they are said to win), or if they go bust (in which case they lose). If the gambler starts with a given stake, what is the probability that they will win the game? We are going to determine this by a *conditioning* argument.

Conditioning is a powerful tool in probability theory. It is based on the following idea. Let A and B denote two events. What is the probability that A will occur if we know in advance that B has occurred? The probability is referred to as a conditional probability (the probability of A given B) and is denoted $P(A/B)$. A little thought shows that

$$P(A/B) = \frac{P(A \cap B)}{P(B)}. \tag{8.7}$$

Note that this formula jives with the definition of independent events given earlier. If A and B are independent events, intuitively one should have $P(A/B) = P(A)$. In view of (8.7) this implies

$$P(A) = \frac{P(A \cap B)}{P(B)},$$

i.e., $P(A \cap B) = P(A)P(B)$, as defined earlier.

As an example in the use of conditional probability, consider the following problem. Suppose we draw two cards from a standard deck without replacement. What is the probability that the first card drawn is a King and the second card is a Queen? Call this event E. The easiest way to calculate $P(E)$ is as follows. The conditional probability of drawing a Queen as the second card *given* that a King was drawn as the first card is $4/51$ (since, after the first card is drawn there are now 51 cards left in the deck and 4 of them are Queens). By (8.7), $P(E) = P(\text{the second card is a Queen/ the first card is a King}) \times P(\text{the first card is a King}) = 4/51 \times 1/4 = 1/51$.

We will need the following elementary result.

Theorem 8.3. *Suppose B_1 and B_2 are events such $B_1 \cap B_2 = \phi$, $B_1 \cup B_2 = S$ (the whole sample space.) Then for any event A, we have*

$$P(A) = P(A/B_1)P(B_1) + P(A/B_2)P(B_2).$$

Proof.

$$\begin{aligned} &P(A/B_1)P(B_1) + P(A/B_2)P(B_2) = P(A \cap B_1) + P(A \cap B_2) \\ &= P\big((A \cap B_1) \cup (A \cap B_2)\big) = P\big(A \cap (B_1 \cup B_2)\big) \\ &= P(A \cap S) = P(A). \end{aligned}$$

□

Returning to the gambler's ruin, let P_m denote the probability that a gambler starting with stake m, will reach N before going bankrupt. The trick is to *condition on the first play of the game.* If the gambler wins the first play, they will then have a stake of $m+1$ and the game proceeds as if started from this position. That is, their probability of eventually winning is P_{m+1}. An analogous situation holds if the gambler loses the first play. By Theorem 8.3, we have

$$\begin{aligned} P_m &= P(\text{eventually reach } N/\text{first play is a win})P(\text{first play is a win}) \\ &\quad + P(\text{eventually reach } N/\text{first play is a loss})P(\text{first play is a loss}) \\ &= P_{m+1} \cdot p + P_{m-1} \cdot q. \end{aligned}$$

This is, provided $1 \leq m \leq N-1$. Clearly, $P_0 = 0$ and $P_N = 1$. This is an example of a *recurrence relation* with boundary conditions. Since $p+q=1$, we may rewrite the above relation as

$$(p+q)P_m = pP_{m+1} + qP_{m-1}.$$

This equation is easily rearranged into the form

$$P_{m+1} - P_m = \frac{q}{p}(P_m - P_{m-1}).$$

Applying this formula successively with $m = 1, 2, 3, \ldots$, yields, for $m = 1, 2, \ldots N$

$$\begin{aligned} P_2 - P_1 &= \frac{q}{p}(P_1 - P_0) = \frac{q}{p}P_1 \\ P_3 - P_2 &= \frac{q}{p}(P_2 - P_1) = \left(\frac{q}{p}\right)^2 P_1 \\ P_4 - P_3 &= \frac{q}{p}(P_3 - P_2) = \left(\frac{q}{p}\right)^3 P_1 \\ &\vdots \\ P_m - P_{m-1} &= \left(\frac{q}{p}\right)^{m-1} P_1. \end{aligned}$$

Adding these equations and noting the cancellations on the left-hand side, we have

$$P_m - P_1 = \frac{q}{p}P_1 + \left(\frac{q}{p}\right)^2 P_1 + \cdots + \left(\frac{q}{p}\right)^{m-1} P_1.$$

That is,

$$P_m = P_1 + \frac{q}{p}P_1 + \left(\frac{q}{p}\right)^2 P_1 + \cdots + \left(\frac{q}{p}\right)^{m-1} P_1. \tag{8.8}$$

In the case $q/p = 1$, i.e., $p = q = .5$, (8.8) implies

$$P_m = mP_1.$$

In the case $q/p \neq 1$, summing the right-hand side of (8.8) by the geometric formula yields

$$P_m = P_1\left(\frac{1-(q/p)^m}{1-(q/p)}\right).$$

Finally, using the condition $P_N = 1$, we may solve for P_1 in each case to obtain

$$P_m = \begin{cases} \dfrac{m}{N}, & \text{if } p = q = .5, \\ \\ \dfrac{1-(q/p)^m}{1-(q/p)^N}, & \text{otherwise.} \end{cases} \tag{8.9}$$

Let's work an example. Suppose the win probability p on each play is 0.48. Our gambler starts, say with \$50 and decides they will quit if they bust or make it to \$100 (the type of strategy most casual gamblers probably employ, if they are sensible!). Formula (8.9) with $p = .48$ and $N = 100$ yields $P_{50} = .018$. Think about this. If you were to bet your entire \$50 stake on a single double or nothing wager with these odds, your probability of ending up with \$100 would be 0.48. Playing the Gambler's ruin, your chance of success is a little less than 1/50. This is why they call it the gambler's ruin and not "the gambler's fast track to fortune."

Exercise 8.7.* *Suppose you play the Gamblers Ruin starting with an initial stake of \$100 and seeking to double to \$200. Calculate the probability of success.*

8.4 RANDOM WALKS

Consider a variation of the Gambler's Ruin, where you have no impulse control which causes you to quit the game if you have won a predesignated amount and where the house will advance you any amount of credit. Your holdings X_n after the nth play of the game then wanders among the set of integers $\{\ldots,-2,-1,0,1,2,\ldots\}$ in a style known as a random walk.

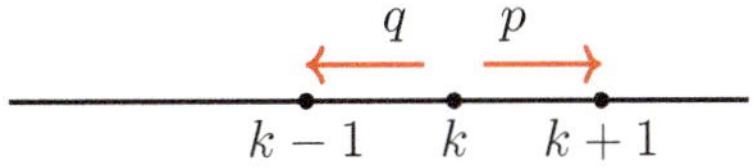

FIGURE 8.3 A random walk on the integers

Suppose $X_0 = 0$. Then we can write

$$X_n = \sum_{k=0}^{n} Z_k, \tag{8.10}$$

where the Z_k are independent Bernoulli random variables taking value 1 with probability p, and -1 with probability $1-p$ (representing a gain or a loss on the kth play).

Exercise 8.8* *Determine* $P(X_{10}=4)$ *in the case when* $p=0.4$ *and* $q=0.6$.

We are interested in the following question: having started off in state 0, is the random walk certain to return to 0 after some finite number of steps, or is there a definite chance that X_n will wander off and never return there? The answer, as we will shortly see, depends on the value of p.

We start by introducing a notation for the return probability to 0. Let

$$f = P(X_n = 0 \text{ for some } n \geq 1/X_0 = 0).$$

If $f=1$, then the random walk is said to be *recurrent* and if $f<1$, then it is said to be *transient.*

Note that if X_n returns to 0, then because of the independence of the steps, the walk starts out afresh, so to speak. So, in the recurrent case, the process is sure to return to state i again, and then again, infinitely many times. In the transient case, the number N of returns to 0 will be finite. The event $\{N=n\}$ represents n successive returns to 0 followed by a non-return, and these happen independently and with respective probabilities f and $1-f$. Thus

$$P(N=n) = f^n(1-f),\ n=0,1,2,\dots,$$

i.e., N is a geometric random variable with parameter $1-f$.

Let P_n denote the probability of returning to 0 in n steps.

Theorem 8.4. *The random walk is transient if the sum*

$$\sum_{n=1}^{\infty} P_n$$

is convergent, and recurrent if the series is divergent.[2]

Note that return to state 0 in n steps is possible only if n is even, requiring an equal number of steps to left and right. The probability of the walk taking k steps to the left and k to the right in $2k$ steps is the binomial probability $\binom{2k}{k}p^k q^k$. Hence the sum in Theorem 8.4, whose convergence or divergence we need to determine, is

$$\sum_{k=1}^{\infty} \binom{2k}{k} p^k q^k. \tag{8.11}$$

We make use of the following result, known as *Stirling's formula* to analyze this series

$$n! \sim n^{n+1/2} e^{-n} \sqrt{2\pi}.$$

[2]See page 168 for the definitions of convergent and divergent series.

where, for two sequences a_n and b_n, the condition $a_n \sim b_n$ means

$$\lim_{n\to\infty} \frac{a_n}{b_n} = 1.$$

Also:

Theorem 8.5. *(i) Suppose $a_n \sim b_n$. Then the series $\sum_n a_n$ and $\sum_n b_n$ either both converge or both diverge.*

(ii) The series

$$\sum_{n=1}^{\infty} \frac{1}{n^p}$$

converges if $p > 1$ and diverges if ≤ 1.

By Stirling's formula, we have

$$\binom{2k}{k} = \frac{2k!}{(k!)^2} \sim \frac{(2k)^{2k+1/2}e^{-2k}\sqrt{2\pi}}{\left(k^{k+1/2}e^{-k}\sqrt{2\pi}\right)^2} = \frac{4^k}{\sqrt{\pi k}}. \tag{8.12}$$

There are two cases to consider. First, suppose $p = q = 1/2$ (the *symmetric* case) in (8.11). Then by (8.12)

$$\binom{2k}{k} p^k q^k \sim \frac{1}{\sqrt{\pi k}} \tag{8.13}$$

and by Theorem 8.5(ii) the series

$$\sum_{k=1}^{\infty} \frac{1}{\sqrt{\pi k}} = \frac{1}{\sqrt{\pi}} \sum_{k=1}^{\infty} \frac{1}{\sqrt{k}}$$

is divergent. It follows from Theorem 8.5(i) and Theorem 8.4 that the random walk is recurrent.

Suppose, on the other hand $p \neq 1/2$. It is an easy exercise in calculus to show that $pq < 1/4$.

Exercise 8.9.⋆ *Prove the last statement.*

By (8.12)

$$\binom{2k}{k} p^k q^k \sim \frac{(4pq)^k}{\sqrt{\pi k}} < (4pq)^k$$

and because $4pq < 1$,

$$\sum_{k=1}^{\infty} (4pq)^k$$

is a convergent geometric series, It follows from Theorem 8.5(i) and Theorem 8.4 that in this case the random walk is transient.

Random walks in higher dimensions

We consider a random walk in two-dimensions. Here, the walk takes place on the lattice of integer pairs (i, j) and at each stage can move either left, right up, or down (see figure 8.4).

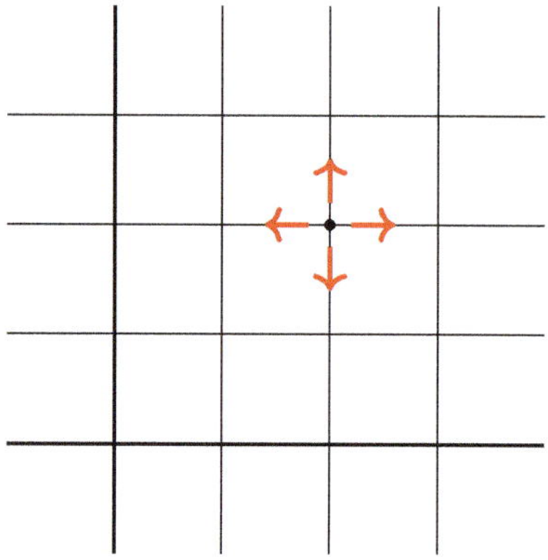

FIGURE 8.4 A random walk on the integer lattice

In two-dimensions, the random walk has a tendency to be more transient than in one-dimension (there are more routes to get lost), so it is safe to assume it will be transient except possibly in the symmetric case where the probability of moving in each of the four directions is $1/4$. This case, which we now analyze, is more delicate.

Assume we are in state $\mathbf{0} = (0, 0)$, and consider the probability P_n of returning to this state in n moves. Again, it is possible only if n is even, so let $n = 2k$. Suppose i of the moves are to the right and j moves are upwards. To return to $\mathbf{0}$, there must also be i moves down and j moves left. We therefore have $2i + 2j = 2k \implies i = k - j$. We decompose the event of starting at $\mathbf{0}$ and returning there in $2k$ moves into these scenarios, i.e., j moves left and right and $k - j$ moves up and down (with $j = 0, 1, 2, \ldots k$). The probability of this event is found from the multinomial distribution (a generalized version of the binomial distribution). The result is

$$\sum_{n=1}^{\infty} nP_n = \sum_{k=1}^{\infty}\sum_{j=0}^{k} \frac{(2k)!}{j!j!(k-j)!(k-j)!}\left(\frac{1}{4}\right)^{2k} = \sum_{k=1}^{\infty}\binom{2k}{k}^2\left(\frac{1}{4}\right)^{2k}.$$

Another calculation with Stirling's formula shows that this sum is asymptotic to

$$\binom{2k}{k}^2\left(\frac{1}{4}\right)^{2k} \sim \frac{1}{\pi k}$$

and if follows from the $p = 1$ case of Theorem 8.5(ii) that the symmetric random walk in two dimensions is also recurrent.

But only just so! The series on which this conclusion is based, $\sum_{k=1}^{\infty} 1/k$, known as the *harmonic* series, is on the borderline between convergence and divergence, with the partial sums $\sum_{k=1}^{n} 1/k$ diverging to infinity, but very, very slowly. The analogous calculation in three-dimensions (where the possible choices of direction at each step are left, right, back, forth, up, down, each with probability 1/8), turns up the series

$$\sum_{k} \frac{1}{k^{3/2}}$$

and this series is *convergent.* So the symmetric random walk in three-dimensions is transient.

In more colorful terms, a drunkard stumbling around on good old two-dimensional mother earth is bound to return home eventually, but a drunken helicopter pilot runs the risk of getting lost forever. (Of course, the pilot might experience a worse problem than transience!)

8.5 CONTINUOUS PROBABILITY

Probability Density Functions

The random variables discussed so far are of a discrete type, taking values in the integers or a finite set. By contrast, a wide range of random phenomena produce outcomes in a continuous spectrum. Modeling such phenomena requires a different set of tools and these come from calculus.

We say that X is a *continuous random variable* if there exists a non-negative integrable function f (the *probability density function*[3] of f) such that

$$P(a < X < b) = \int_a^b f(x)dx.$$

Thus, the probability that X falls between a and b is being specified as the area lying under the graph $y = f(x)$ between $x = a$ and $x = b$.

As an example, chose a point at random from the interval $[a, b]$ and denote it by X. Then X is a continuous random variable with pdf

$$f(x) = \begin{cases} \dfrac{1}{b-a} & \text{if } a \le x \le b, \\ 0 & \text{otherwise.} \end{cases} \tag{8.14}$$

[3]Implicit in this definition is the property

$$\int_{-\infty}^{\infty} f(x)\, dx = 1.$$

The fact that f is constant on the interval $[a,b]$ indicates the complete randomness of the choice. (No one point is "more likely" to be chosen from the interval than any other point.) This is called a *Uniform* distribution and denoted by $U(a,b)$.

The most important continuous probability distribution in statistical applications is the *normal*, (or *Gaussian*) distribution. (Yet another mathematical object named for Gauss!) The pdf for a normal distribution has the form

$$f(x) = \frac{1}{\sqrt{2\pi}\sigma} e^{-\dfrac{(x-\mu)^2}{2\sigma^2}}, \tag{8.15}$$

where the parameters μ and σ denote the mean and standard deviation of the distribution (to be defined below.) The case with $\mu = 0$ and $\sigma = 1$ is said to be a *standard normal* distribution.

The graph of a normal distribution has the characteristic bell-shape, with which anybody who has taken an elementary statistics course is bound to be familiar. Some examples are shown in the figure below. An important feature of the curves is the extremely thin tails (the parts of the graph on the left and right). This implies there is a very small probability that a random variable with a normal distribution will fall into a range more than a few standard deviations away from the mean.

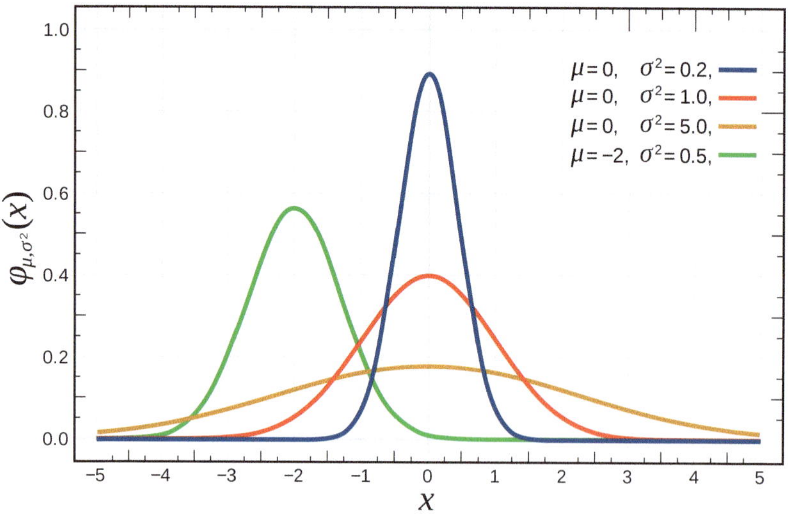

FIGURE 8.5 Normal curves with differing means and variances

The Mean and Variance of a Continuous Random Variable

Let X be a continuous random variable with pdf f. There is a natural analogue of the notion of mean of X previously defined for discrete random variables. It is the integral

$$\int_{-\infty}^{\infty} xf(x)dx.$$

The variance of X is defined by

$$\int_{-\infty}^{\infty} (x-\mu)^2 f(x)dx.$$

where μ is the mean of X.

As an example, we calculate the mean of the normal distribution (8.15). That is,

$$\frac{1}{\sqrt{2\pi}\sigma}\int_{-\infty}^{\infty} xe^{-\dfrac{(x-\mu)^2}{2\sigma^2}}\,dx.$$

The change of variable

$$t = \frac{x-\mu}{\sigma}$$

transforms the integral to

$$\begin{aligned}
&\frac{1}{\sqrt{2\pi}}\int_{-\infty}^{\infty} (\sigma t+\mu)e^{-t^2/2}\,dt \\
&= \mu\cdot\frac{1}{\sqrt{2\pi}}\int_{-\infty}^{\infty} e^{-t^2/2}\,dt + \frac{\sigma}{\sqrt{2\pi}}\int_{-\infty}^{\infty} te^{-t^2/2}\,dt \\
&= \mu
\end{aligned}$$

since the integral in the second term is 0, being the integral of an odd function over a symmetric interval about 0. We have used the fact that the standard normal density integrates to 1 to evaluate the first term.

Exercise 8.10. *Show, by a similar argument that the variance of the distribution is* σ^2.

Exercise 8.11. *Show that the* $U(a,b)$ *distribution has mean*

$$\frac{a+b}{2} \tag{8.16}$$

and variance

$$\frac{(b-a)^2}{12}. \tag{8.17}$$

8.6 PROBABILISTIC LIMIT LAWS

Limit laws are among the most important results in probability theory. To get a feel for the topic, consider a sequence of repetitions of the experiment of choosing a point at random from the interval $[0, 1]$. The chosen numbers $X_1, X_2, X, \ldots$ comprise a sequence of independent random variables with a uniform distribution on the interval $[0, 1]$. By (8.16) and (8.17), the mean and variance of the individual X_i's are, respectively, $1/2$ and $1/12$.

Consider the distribution of the *sample mean* of the first n of these random variables, i.e.,

$$\bar{X}_n = \frac{1}{n}\sum_{k=1}^{n} X_k.$$

By Theorem 8.1[4]

$$E[\bar{X}_n] = \frac{1}{n}\sum_{k=1}^{n} E[X_k] = \frac{1}{n}\sum_{k=1}^{n} \mu = \frac{1}{2} \tag{8.18}$$

and

$$Var(\bar{X}_n) = \frac{1}{n^2}\sum_{k=1}^{n} Var[X_k] = \frac{1}{n^2}\sum_{k=1}^{n} \sigma^2 = \frac{1}{12n}. \tag{8.19}$$

Thus we see, e.g., that $\bar{X}_{12}$ has mean $1/2$ and standard deviation $1/12$. Chebyshev's inequality yields the estimate

$$P(.25 < \bar{X}_{12} < .75) \geq 8/9.$$

In particular, $\bar{X}_{12}$ does not itself have a uniform distribution on $[0,\ 1]$, or the probability would be 0.5. Furthermore, as we let n become larger, by (8.19), $Var(\bar{X}_n)$ grows smaller, so, again by Chebyshev, the probability that $\bar{X}_n$ lies in any interval $(1/2-\epsilon, 1/2+\epsilon)$ centered at the mean, will approach ever closer to 1. This tendency for sample means to "cluster" around the theoretical mean of the distribution, is expressed by the following result, known as the weak law of large numbers (WLLN).

Theorem 8.6. *Let X_n be sequence of random variables with mean μ and finite variance. For all $\epsilon > 0$*

$$P\big(|\bar{X}_n - \mu| > \epsilon\big) \to 0, \ \text{as } n \to \infty.$$

Exercise 8.12. *Prove Theorem 8.6 using Chebyshev's inequality.*

The next result is known as the central limit theorem (CLT).

[4]The theorem was stated for discrete random variables but also holds in the continuous case, as does Chebyshev's inequality.

Theorem 8.7. *Suppose X_n is a sequence of IID (independent, identically distributed) random variables with mean μ and standard deviation σ. Define*[5]

$$Z_n = \frac{\bar{X}_n - \mu}{\sigma/\sqrt{n}}.$$

Then for all $a < b$,

$$P(a < Z_n < b) \to \frac{1}{\sqrt{2\pi}} \int_a^b e^{-x^2/2}\, dx, \quad \text{as } n \to \infty. \tag{8.20}$$

Note that the integral in (8.20) is the probability associated with the standard normal distribution. We say Z_n *converges in distribution* to standard normal. Remarkably, this holds irrespective of the original distribution of the observations. This is the reason the Gaussian distribution is referred to as "normal".

Owing to its universal nature, CLT is of fundamental importance in statistical applications. Here is an example.

A soft drinks company asserts that the average contents of their 2 liter bottles is, (naturally) 2 liters and has a standard deviation of 0.05 liters. A consumer organization asserts that the company is providing short measure. A sample of 100 bottles is examined and found to have an average of 1.982 liters. Slightly lower than the promised 2 liters, but might not such a small discrepancy just be due to "sampling" error?

This question can be addressed by a sort of statistical "proof by contradiction." Assume the company's claim is true. Then the sample has been drawn from a population with mean 2.0 and standard deviation 0.05. Since the sample size of 100 is relatively large, CLT allows us to assume that the sample mean $\bar{X}$ follows a normal distribution. The asserted mean of $\bar{X}$ is 2.0 and its standard deviation is $.05/\sqrt{100} = 0.005$. Consulting a table of normal probabilities shows that the likelihood of $\bar{X}$ returning the observed value of 1.982 (a value more than 3 standard deviations lower than the mean) is less than 1/1000. We conclude that the test provides strong evidence to support the counter claim.

Both WLLN and CLT concern the long term behavior of the *distribution* of the sample mean of a sequence of IID random variables. By contrast, the strong law of large numbers (SLLN) which we now present, address the convergence of the random variables themselves.

The strong law is founded on the notion of *almost sure* convergence. We say that an event E happens "almost surely" (denoted *a.s.*) if E happens with probability 1. For example, suppose I am asked to choose a number X at random between 0 and 1. Then $P(X = 0.5) = 0$, as we saw earlier in the

[5]Note that, in subtracting μ and dividing by $\sigma/\sqrt{n}$, we have "standardized" the Z_n so that they have mean 0 and variance 1.

discussion of the uniform distribution. Hence the event $\{X \neq 0.5\}$ happens with probability $1 - 0 = 1$, i.e., almost surely. In mathematical statements concerning continuous probability, almost sure convergence is usually the best one can do. (There is a standing joke that probabilists are never sure about anything, they are only ever *almost* sure.)

The strong law is as follows.

Theorem 8.8. *Suppose that X_n is a sequence of IID random variables with mean μ and finite variance. Then*

$$\frac{X_1 + X_2 + \cdots + X_n}{n} \to \mu \text{ as } n \to \infty, \quad a.s.$$

As an application of the strong law, we offer the following probabilistic method to estimate the number π. Suppose a *very bad* darts player throws a million darts at a board ... okay, if this scenario sounds a little implausible, then use a random number generator to generate a million number pairs (x, y) inside the square $[-1, 1] \times [-1, 1]$. Calculate the proportion of *hits*, i.e., those numbers which land inside the circle $x^2 + y^2 \leq 1$. The probability of a hit each time is $\pi/4$ (the ratio of the area of the circle to the area of the square.) By SLLN, the proportion of hits will give a very accurate approximation to $\pi/4$.

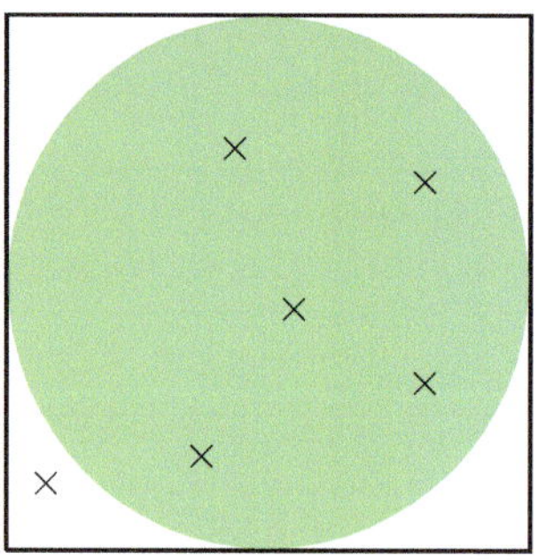

FIGURE 8.6

Exercise 8.13.[⋆] *Use SLLN to give another proof that the asymmetric random walk (say with $p > q$) is recurrent.*

8.7 THE NORMAL NUMBER THEOREM

In 1909, the French Mathematician Émile Borel used probabilistic methods to prove a remarkable theorem about real numbers, ushering in a new approach to number theory. Probabilistic number theory was to become a major area of research in the second half of the twentieth century with the pioneering work of Paul Erdős and others. Borel's foundational contribution is the subject of this section.

We begin by introducing the leading figures (pun intended) in the drama. A number $x \in [0, 1]$ is said to be *normal in base* $d \geq 2$ if, in the representation of the number in base b, each of the digits $0, 1, \ldots d-1$ occurs with equal frequency $1/b$. An example is the number which, when expressed in base 5 is

$$0.\overline{01234} = 0.012340123401234\ldots$$

This number is normal in base 5.

x is said to be *absolutely normal* (or just *normal*) if x is normal *in every base* $b \geq 2$.

No *rational number* p/q has this property for the obvious reason that, in base q,

$$p/q = 0.p0000\ldots,$$

so is very definitely non-normal in base q. It is hard to come up with even a single example of a normal number.[6] It is conjectured that natural irrationals such as $\sqrt{2}, e$ and π are normal, but this has not been proved.

The normal number theorem is as follows.

Theorem 8.9. *Almost all numbers in* $[0, 1]$ *are normal.*

The term *almost all* in the statement of the theorem has the same meaning as almost surely in the previous section; it means the event happens everywhere except on a set of zero probability. In this setting, the role of probability is played by the so-called *Lebesgue measure.* Lebesgue measure (denoted by λ) can be thought of as a sort of generalized version of length, so in particular

$$\lambda\big([a, b]\big) = b - a$$

for any interval $[a, b] \subset [0, 1]$.

If $A_1, A_2, \ldots$ is any (finite or countably infinite) collection of sets such that $\lambda(A_j) = 0$, for all j, then[7]

$$\lambda(A_1 \cup A_2 \cup \ldots) = 0. \tag{8.21}$$

We now outline Borel's proof of the normal number theorem. It is based on the strong law of large numbers (SLLN). We first indicate how it can be proved that almost all numbers in $[0, 1]$ are normal to base 2.

[6] A method does, in fact, exist to construct normal numbers but these numbers are very sparse.

[7] This follows from the property: $\lambda\Big(\bigcup_k A_k\Big) \leq \sum_k \lambda(A_k)$.

FIGURE 8.7 Émile Borel

To this end we define a sequence of functions (random variables) X_n on $[0,1]$ by

$$X_n(x) = \begin{cases} 1, & \text{if } x_n = 0, \\ 0, & \text{if } x_n = 1, \end{cases}$$

where x_n is the nth digit in the binary representation

$$0.x_1x_2x_2\ldots$$

of x.

For example, 0.75 has the binary representation $0.11000\ldots$, so

$$X_1(0.75) = 1,$$
$$X_2(0.75) = 1,$$
$$X_n(0.75) = 0, \text{ for } n \geq 3.$$

It is straightforward to show that $X_1, X_2, \ldots$ are all independent, and have identical Bernoulli distributions with $p = 1/2$, hence mean $1/2$. It follows from SLLN that, outside of a set of Lebesgue measure zero

$$\frac{X_I(x) + \ldots X_n(x)}{n} \to 1/2, \text{ as } n \to \infty. \tag{8.22}$$

FIGURE 8.8 Paul Erdős (author Kmhkmh)

Since the sum $X_I(x) + \ldots X_n(x)$ gives the number of 0's in the first n places of the binary representation of x, (8.22) asserts that half of the digits in x are 0's and half are 1's, i.e., x is normal to base 2. Let B_2 denote the subset of $[0, 1]$ on which (8.22) does *not* happen, so $\lambda(B_2) = 0$. Employing a double negative, we have shown that outside B_2, all the numbers in $[0, 1]$ are normal to base 2.

A similar argument serves to prove that the same result for base 3, although we should point out that the argument here requires an additional step. It is necessary to introduce *two* sets of Bernoulli random variables which count the numbers of 0's and the number of 1's in the base-3 representation of x:

$$X_n(x) = \begin{cases} 1, & \text{if } x_n = 0, \\ \\ 0, & \text{if } x_n = 1 \text{ or } 2, \end{cases}$$

$$Y_n(x) = \begin{cases} 1, & \text{if } x_n = 1, \\ \\ 0, & \text{if } x_n = 0 \text{ or } 2. \end{cases}$$

Repeating the previous argument with both the X's, and the Ys we find that for x outside of a set B_3 with $\lambda(B_3) = 0$, x is normal to base 3.

Continuing in this manner, we obtain a sequence of sets B_d $(d \geq 2)$ such that $\lambda(B_d) = 0$ and if $x \notin B_d$, then x is normal to base d.

Define

$$B = \bigcup_{b=1}^{\infty} B_b.$$

By (8.21), $\lambda(B) = 0$. If $x \notin B$ then $x \notin B_d$ for all d, which implies x is normal to base d for all $(d \geq 2)$, i.e., x is normal. The proof is complete.

Suggestions for Further Reading

Sheldon M. Ross. *Introduction to Probability Models.* Academic Press, 13th edition, 2023.

John W. Lamperti. Probability: A Survey of the Mathematical Theory. Wiley, 2nd edition, 1996.

CHAPTER 9

Countability and Computability

This chapter starts with some stunning results of Georg Cantor concerning infinite sets. He defined *cardinality*, a measure of the size of a set. For finite sets, the cardinality is just the number of elements in the set, but Cantor's definition extends this notion of size to infinite sets. He proved that the cardinality of the integers is equal to the cardinality of the rational numbers but less than the cardinality of the real numbers. His major theorem shows how given any set, we can find a set with greater cardinality—proving that there are an infinite number of infinities!

After looking at Cantor's work done in the late nineteenth century, we turn to the work of David Hilbert. At the turn of the twentieth century, Hilbert listed twenty three unsolved problems he considered the most important. Hilbert was the leading mathematician of his day, and many mathematicians and logicians started work on trying to solve them. The solutions to some were totally unexpected!

Hilbert wanted to put mathematics on a sound footing. He felt mathematics should start with axioms and rules of inference. Any theorem should be able to be proved by starting with axioms and applying the rules of inference. Kurt Gödel's incompleteness theorems showed that Hilbert's approach was doomed.

Another of Hilbert's problems concerned a question that Cantor could not answer. This is called the *continuum hypothesis*: Is there a subset of the real numbers with cardinality greater than the integers but less than the reals? Again, the answer is surprising. Given our standard axioms for set theory, the continuum hypothesis can be neither proved nor disproved.

Alan Turing tackled another problem of Hilbert's, though not one of the twenty three. To begin, he needed to give a definition of *algorithm*. To do this,

DOI: 10.1201/9781003470915-9

he constructed what is now known as, a Turing machine. We look at Turing's work and the connection to the theory of computation.

We then turn to one of the outstanding unsolved questions in computer science: Does P equal NP? The chapter concludes with an application of internet security, RSA encryption, that depends on a negative answer to this question.

9.1 INFINITY

Aristotle distinguished between two types of infinity: *potential infinity* describes a process that is ongoing and never completed; *actual infinity* is used for an infinite number of things. Actual infinity is now generally accepted—we talk of the set of natural numbers, for example—but this is a fairly recent phenomenon. Gauss did not think that mathematics needed actual infinities, writing in a letter:

> I protest against the use of infinite magnitude as something completed, which is never possible in mathematics. Infinity is merely a way of speaking, the true meaning being a limit which certain ratios approach indefinitely close, while others are permitted to increase without restriction.

Potential infinities are clearly a part of mathematics. For example, we write:

$$\sum_{k=0}^{\infty} \left(\frac{1}{2}\right)^k = 2.$$

To formally prove a statement like this, you need to show for any $\varepsilon > 0$ there is a number N with the property

$$2 - \varepsilon < \sum_{k=0}^{n} \left(\frac{1}{2}\right)^k < 2 + \varepsilon, \qquad \text{whenever } n \geq N.$$

This is saying for any tolerance about 2, the finite sum will be within the tolerance for large enough values of n.

Calculus involves limits as "n goes to infinity," but all of these are potential infinities. Most mathematicians up until the second half of the nineteenth century agreed with Gauss: actual infinites are not part of mathematics. Georg Cantor's work would change this.

One-to-One and Onto Functions

Suppose we are given a function f that maps elements of a set X to elements of a set Y. We say the function is *one-to-one* if it sends distinct elements of X to distinct elements of Y. More formally,

Definition 9.1. *f is* one-to-one *if whenever $x_1 \neq x_2$, then $f(x_1) \neq f(x_2)$. Equivalently: f is* one-to-one *if whenever $f(x_1) = f(x_2)$, then $x_1 = x_2$.*

Definition 9.2. *A function f from X to Y is* onto *if given any element y of Y, there is an x in X with $f(x) = y$.*

A function that is both one-to-one and onto is called a *bijection.* A bijection from a set X to a set Y pairs the elements of X and Y so that each element of X has its unique partner in Y, and each element of Y has its unique partner in X. We say that the elements of X can be put into a *one-to-one correspondence* with elements of Y.

As an example, consider $f : \mathbb{R} \to \mathbb{R}$ defined by $f(x) = 2x$. This function is both one-to-one and onto, so a bijection. However, the function $f : \mathbb{Z} \to \mathbb{Z}$ defined by $f(x) = 2x$ is one-to-one, but not onto—nothing is sent to 3, for example—so it is not bijection.

Cardinality

Cantor wanted a way of saying two sets had the same size. Finite sets A and B have the same size if they have the same number of elements. When we count the number of elements in a set, we are giving a bijection from a subset of the natural numbers to the set in question: A set A has n elements if and only if there is a bijection between $\{1, 2, 3, \ldots, n\}$ and A.

Cantor extended this idea to sets in general.

Definition 9.3. *Sets A and B have the same* cardinality *if there exists a bijection between them. We will denote the cardinality of A by $|A|$.*

Two finite sets have the same cardinality if they have the same number of elements, so the cardinality of the set is just this number. However, Cantor defined cardinality in order that he could also talk about the size of infinite sets.

We let $\mathbb{N}$ denote the set of natural numbers and $2\mathbb{N}$ the set of even natural numbers. Let

$$f : \mathbb{N} \to 2\mathbb{N} \text{ be defined by } f(n) = 2n.$$

If $2m = 2n$, then $m = n$, showing f is one-to-one. It is also onto. Given any even natural number $2k$, the natural number k gets mapped to it. Consequently, we have a bijection between $\mathbb{N}$ and $2\mathbb{N}$, which means they have the same cardinality. Cantor used the Hebrew letter aleph to denote cardinals with $\aleph_0$ for the cardinality of the natural numbers.

$$|\mathbb{N}| = |2\mathbb{N}| = \aleph_0.$$

FIGURE 9.1 Georg Cantor

The even natural numbers have the form $2k$ for $k = 1, 2, \ldots$. The odd natural numbers have the form $2k+1$ for $k = 0, 1, 2, \ldots$. We can define a function from the natural numbers to the integers, denoted $\mathbb{Z}$, by:

$$f(2k) = k, \quad f(2k+1) = -k.$$

We have

$$f(1) = 0,\ f(2) = 1,\ f(3) = -1,\ f(4) = 2,\ f(5) = -2, \ldots$$

It is straightforward to check that f is a bijection from the natural numbers to the integers, telling us

$$|\mathbb{N}| = |\mathbb{Z}| = \aleph_0.$$

The property that a set X can be put into a one-to-one correspondence with a proper subset of X seems strange when we first meet it because it is not true for finite sets, and these are the ones we have experience with. A proper subset of a finite set has fewer elements than the whole set. But *any* infinite set X can be put into one to a one-to-one correspondence with a proper subset of itself. We can take this as a defining property of infinite sets: A set is infinite if and only if it can be put into a one-to-one correspondence with a proper subset.[1]

[1]This definition was first given by Dedekind in 1872.

Suppose we have a one-to-one function from a set X to a set Y. Then there must be a one-to-one correspondence between X and some subset of Y (possibly Y, itself). So we can say $|X| \geq |Y|$. Now suppose there is also a one-to-one function from Y to X, telling us $|Y| \geq |X|$. Can we conclude that $|X| = |Y|$? Cantor certainly believed this to be true, but didn't give a proof. This is not as obvious as it might first seem. To prove this, you need to construct a bijection from the two one-to-one functions: one going from X to Y and the other going from Y to X. Fortunately, this construction is possible. This result is now named after Felix Bernstein and Ernst Schröder who independently gave proofs.

Theorem 9.1. *(Schröder–Bernstein) Suppose $f : X \to Y$ and $g : Y \to X$ are both one-to-one functions, then there exists a bijection from X to Y.*

This can be restated as the following useful property of inequalities:

Corollary 9.1. *Let X and Y denote two sets. If $|X| \leq |Y|$ and $|Y| \leq |X|$, then $|X| = |Y|$.*

Cardinality of the Rational Numbers

We will use the Schröder–Bernstein theorem to determine the cardinality of the rational numbers.

First, we note that any rational number, with the exception of 0, can be written uniquely in the form p/q, where p and q are coprime and $q \geq 1$. The one exception, 0 will be written as $0/1$. We assume our numbers are written in this form.

Then we define a function from the rationals, denoted by $\mathbb{Q}$, to the natural numbers in the following way:

1. $f(0) = 1$.
2. If $p \geq 0$, then $f(p/q) = 2^p 3^q$.
3. If $p < 0$, then $f(p/q) = 2^p 3^q 5$.

There is a 5 in the prime factorization of $f(p/q)$ if and only if p/q is negative. The exponent of the powers of two in the prime factorization of $f(p/q)$ gives p. The exponent of the powers of three in the prime factorization of $f(p/q)$ gives q.

It is clear that distinct rational numbers get sent to distinct natural numbers, so the function is one-to-one.

Since $f : \mathbb{Q} \to \mathbb{N}$ is one-to-one, we have $|\mathbb{Q}| \leq |\mathbb{N}|$. We can define a one-to-one function $g : \mathbb{N} \to \mathbb{Q}$ by $g(n) = n$, telling us the less surprising result: $|\mathbb{N}| \leq |\mathbb{Q}|$. Schröder–Bernstein then gives $|\mathbb{Q}| = |\mathbb{N}|$.

So far, we have shown

$$|2\mathbb{N}| = |\mathbb{N}| = |\mathbb{Z}| = |\mathbb{Q}| = \aleph_0.$$

Cantor then showed that the real numbers do not have cardinality $\aleph_0$. They have a larger cardinality—a stunning result showing there is more than one type of infinity.

Cantor's diagonal argument

He considered the real numbers in the unit interval $0 < x < 1$, which we will denote using interval notation as $(0, 1)$. These can be written as decimals with 0 before the decimal point. To prove the cardinality of $(0, 1)$ is not equal to the cardinality of $\mathbb{N}$, you need to show it is impossible to construct a bijection between $(0, 1)$ and $\mathbb{N}$. Cantor did this using a proof by contradiction. He assumed there was a bijection and then derived a contradiction.

If there is a bijection f from $\mathbb{N}$ to $(0, 1)$, $f(1)$ will be in $(0, 1)$, we let $0.a_{11}a_{12}a_{13}\ldots a_{1n}\ldots$ denote its decimal representation. Similarly, we denote $f(2)$ by $0.a_{21}a_{22}a_{23}\ldots a_{2n}\ldots$, and so on. We can list the natural numbers in one column and the corresponding decimal in another column.

$$\begin{array}{ll}
1 & 0.\mathbf{a_{11}}a_{12}a_{13}\cdots a_{1n}\cdots \\
2 & 0.a_{21}\mathbf{a_{22}}a_{23}\cdots a_{2n}\cdots \\
3 & 0.a_{31}a_{32}\mathbf{a_{33}}\cdots a_{2n}\cdots \\
\vdots & \vdots \\
n & 0.a_{n1}a_{n2}a_{n3}\cdots \mathbf{a_{nn}}\cdots \\
\vdots & \vdots
\end{array}$$

We now construct a decimal number $b = 0.b_1b_2b_3b_4\cdots$ in the following way: choose b_1 so that it does not equal a_{11}, b_2 not equal to a_{22}, b_3 not equal to a_{33}. In general, choose b_n to not be equal to a_{nn}. Looking at the column above, we are choosing the bs to not equal the diagonal elements of the as. Moreover, we can do this in such a way that none of the digits b_n are 0 or 9.

Before we continue, we should look at a technicality. Could b have two different decimal representations? Some numbers do. For example, the number 2 can be written as $2.000000\ldots$ and also as $1.9999999\ldots$. Decimals that end in an infinite string of 9s can be rewritten to have an infinite string of 0s and vice versa. However, these arc thc only numbers with two decimal expansions. We chose b to have none of its digits equal to 0 or 9. Its decimal expansion is unique.

It is clear that b belongs to $(0,1)$. We are assuming $f : \mathbb{N} \to (0,1)$ is a bijection, so b must appear somewhere in the righthand column. But for any n, b cannot be $f(n)$ because their decimal representations differ in the nth place. So b cannot be in the righthand column. We have a contradiction: b is both in and not in the list. Consequently, our initial hypothesis is incorrect. There are no bijections from $\mathbb{N}$ to $(0,1)$.

Since there doesn't exist a bijection, the two sets must have different cardinalities.

$$\aleph_0 = |\mathbb{N}| \neq |(0,1)|.$$

The function $g : \mathbb{N} \to (0,1)$, where $g(n) = 1/n$ is one-to-one, telling us

$$|\mathbb{N}| \leq |(0,1)|.$$

We conclude

$$\aleph_0 = |\mathbb{N}| < |(0,1)|.$$

We have found a set with a larger cardinality. We have found a larger infinity. This is denoted by $\mathfrak{c}$. We often call $\aleph_0$ a countable infinity and call $\mathfrak{c}$ an uncountable infinity.

Bijections can be found between the interval $(0,1)$ and the set of real numbers $\mathbb{R}$, so these two sets have the same cardinality.

Exercise 9.1. *Show* $f : (0,1) \to \mathbb{R}$ *defined by* $f(x) = \tan((x-0.5)\pi))$ *is a bijection.*

Exercise 9.2* *Show* $g : \mathbb{R} \to (0,1)$ *defined by* $g(x) = 1/(1+e^x)$ *is a bijection.*

Exercise 9.3* *Using binary strings, show* $2^{\aleph_0} = |[0,1)| = \mathfrak{c}$.

To summarize, we have:

$$|2\mathbb{N}| = |\mathbb{N}| = |\mathbb{Z}| = |\mathbb{Q}| = \aleph_0 < \mathfrak{c} = |(0,1)| = |\mathbb{R}|.$$

We have found two infinite cardinals $\mathfrak{c}$, the cardinality of the continuum, and $\aleph_0$, the cardinality of countably infinite sets. At this stage, it is natural to ask if there are other infinities. Are there some infinite cardinals that lie between $\aleph_0$ and $\mathfrak{c}$? Cantor believed the answer was no, but could not prove it. As we will see later, there is not a simple *Yes/No* answer. The question of whether there are cardinalities greater than $\mathfrak{c}$ is easier to answer.

Cantor's first attempt to construct a larger cardinal number was to go from the one-dimensional interval $(0,1)$ to the two-dimensional unit square. Clearly the cardinality of the unit square must be at least the cardinality of the unit line. However, Cantor was able to construct a one-to-one function from the unit square to the unit line, showing that the cardinality of the unit square is equal to the cardinality of the unit interval. Figure 9.2 shows the coordinates of a point in the square. Cantor took the coordinates and interleaved them

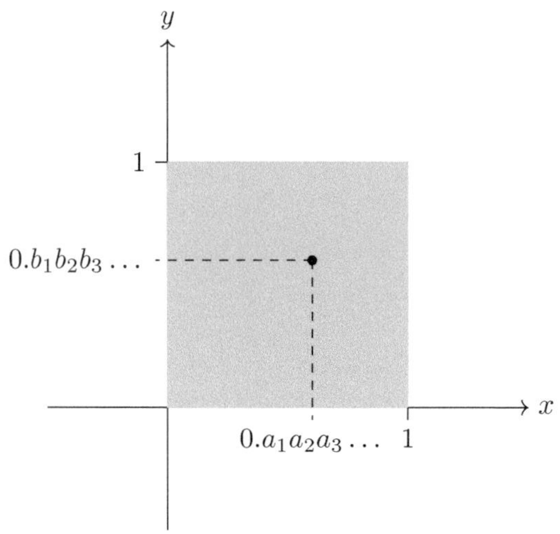

FIGURE 9.2 Unit square with coordinates of a point

to give the function. The point with coordinates $(0.a_1a_2a_3\ldots, \mathbf{0.b_1b_2b_3}\ldots)$ gets sent to $0.a_1\mathbf{b_1}a_2\mathbf{b_2}a_3\mathbf{b_3}\ldots$.

This argument extends to three and higher dimensions. The sets $\mathbb{R}, \mathbb{R}^2, \mathbb{R}^3, \ldots$ all have the same cardinality—$\mathfrak{c}$. Cantor wrote to Dedekind about this surprising result saying, "I see it but I do not believe it!"

Cantor's Theorem

Cantor then looked at all possible subsets of a given set S. For example, the set $S = \{1, 2, 3\}$ has the eight subsets:

$$\phi, \{1\}, \{2\}, \{3\}, \{1,2\}, \{1,3\}, \{2,3\}, \{1,2,3\},$$

where ϕ denotes the empty set—the set with no elements. We also include the set S as a subset of itself.

The set containing all the subsets of a set S is called *the power set* of S and denoted by $\mathcal{P}(S)$. For our example with $S = \{1, 2, 3\}$,

$$\mathcal{P}(S) = \{\phi, \{1\}, \{2\}, \{3\}, \{1,2\}, \{1,3\}, \{2,3\}, \{1,2,3\}\}.$$

Exercise 9.4.* *In the example, $|S| = 3$ and $|\mathcal{P}(S)| = 2^3$. Show for any finite set S that $|\mathcal{P}(S)| = 2^{|S|}$.*

Exercise 9.5. *Show $|\mathcal{P}(\mathbb{N})| = 2^{\aleph_0} = \mathfrak{c}$.*

Cantor found an ingenious argument that shows there can be no bijection between any set S and its power set $\mathcal{P}(S)$. He showed that any function from S to $\mathcal{P}(S)$ cannot be onto.

Theorem 9.2. *(Cantor)*

Given any set S and any function $f : S \to \mathcal{P}(S)$, f is not onto.

Proof. Given a function $f : S \to \mathcal{P}(S)$, there are two possibilities for each $s \in S$, either $s \in f(s)$ or $s \notin f(s)$. Let T denote the set of all elements s that do not belong to $f(s)$.

$$T = \{s \mid s \notin f(s)\}.$$

Clearly, T is a subset of S and consequently an element of $\mathcal{P}(S)$. We want to show that f is not onto. The heart of the proof is to show that no element of S gets mapped to T. If there was an element $t \in S$ with $f(t) = T$, then there are two possibilities, either $t \in T$ or $t \notin T$. We examine both of these and show each leads to a contradiction.

The set T is defined to be the set of elements s such that $s \notin f(s)$. If $t \in f(t)$, then by the definition of T, $t \notin T$, but $t \in f(t) = T$, which gives a contradiction.

If $t \notin f(t)$, then by the definition of T, $t \in T$, but this contradicts $t \notin f(t) = T$. □

Exercise 9.6. *Given any set S, show how to construct a one-to-one function $f : S \to \mathcal{P}(S)$.*

Theorem 9.2 and exercise 9.5 give us one of Cantor's most stunning results.

Theorem 9.3. *For any set S, $|S| < |\mathcal{P}(S)|$.*

Cantor's theorem tells us that $\mathcal{P}(\mathbb{R})$ has cardinality greater than the cardinality of $\mathbb{R}$. Giving us a cardinality greater than $\mathfrak{c}$. We can apply the theorem repeatedly to give a countable infinite number of infinities:

$$\aleph_0 < \mathfrak{c} < |\mathcal{P}(\mathbb{R})| < |\mathcal{P}\left(\mathcal{P}(\mathbb{R})\right)| < |\mathcal{P}\left(\mathcal{P}\left(\mathcal{P}(\mathbb{R})\right)\right)| < \cdots$$

Cantor published his work in 1891. The mathematical community was polarized. Some hailed Cantor as a genius, while others thought the work nonsensical and should not be considered a part of mathematics. One of the leading critics was Leopold Kronecker who actively worked to stop Cantor's work from being published. Kronecker was a powerful mathematician and a professor at the prestigious University of Berlin. Cantor had been a student at this university and always wanted an appointment there. Kronecker's attacks were a crushing disappointment.

Despite the attacks, Cantor continued his work on infinity and sets. He believed that there was no cardinal number strictly between $\aleph_0$ and $\mathfrak{c}$, but could not prove it. This problem became known as *the continuum hypothesis.*

Cantor also considered the ordering of elements in sets. He defined a set to be *well-ordered* if every subset has a least element. The natural numbers $\mathbb{N}$ are well-ordered with the usual ordering. The set of integers $\mathbb{Z}$ and the rational numbers $\mathbb{Q}$ are not well-ordered with their usual orderings but they can be well-ordered. This follows from the fact that there is a bijection from the natural numbers to these sets, so we can reorder them using the ordering on the natural numbers. If

$$f : \mathbb{N} \to S$$

is a bijection, we can define the ordering on S by

$$s < s' \text{ if and only if } f^{-1}(s) < f^{-1}(s').$$

For example, we had a bijection f from the natural numbers to the integers with

$$f(1) = 0,\ f(2) = 1,\ f(3) = -1,\ f(4) = 2,\ f(5) = -2, \ldots.$$

This gives the ordering

$$0, 1, -1, 2, -2, \ldots,$$

which is a well-ordering of the integers.

It is clear that every countable set can be well-ordered, but what about uncountable sets, the reals, for example? Cantor believed that every set could be well-ordered but could not prove it.

Cantor also realized there was a problem concerning how to define sets. For example, if you let U denote the set of all possible sets, then it must be the largest set. However, his theorem says that $\mathcal{P}(U)$ will have a greater cardinality. Cantor believed that this showed there was not a set of all sets, that sets could not be too big, but he never clearly spelled out what the criteria for a set were.

The Cantor Set

We briefly consider an interesting set that is named after Cantor. The set was first discovered by Henry J. S. Smith in 1874, but Cantor mentioned it in 1883. Nowadays, it is referred to as the *Cantor set.*

We construct it from the unit interval $[0, 1]$ by an iterative process in which the middle thirds of intervals are deleted. The first iteration deletes $(1/3, 2/3)$ leaving

$$\left[0, \frac{1}{3}\right] \cup \left[\frac{2}{3}, 1\right].$$

The second iteration deletes the middle thirds of these two intervals leaving

$$\left[0, \frac{1}{9}\right] \cup \left[\frac{2}{9}, \frac{1}{3}\right] \cup \left[\frac{2}{3}, \frac{7}{9}\right] \cup \left[\frac{8}{9}, 1\right].$$

In the third iteration we remove the middle thirds of these four closed intervals, and so on.

The process can also be described by writing the numbers in the interval using base 3 (ternary). Such a number has the form $0.a_1a_2a_3\ldots$, where the digits a_i can be 0, 1, or 2. The first stage of the process keeps all the numbers for which a_1 is either 0 or 2. The second iteration keeps all the remaining numbers with a_2 equal to 0 or 2. The kth iteration keeps all the remaining numbers that with a_k equal to either 0 or 2. After we have finished the process, we are left with the Cantor set. These are the numbers that can be written in base 3 without using the digit 1.

Exercise 9.7. *Show, in base* 3, *that* $0.02222\cdots = 0.1000\ldots$.

Exercise 9.8. *Show that if a number has two ternary expansions, then it belongs to the Cantor set.*

(This is the reason we described the process as "keeping all the numbers with a_k *equal to either* 0 *or* 2*" and* not *saying "delete all numbers with* a_k *equal to* 1.*")*

Exercise 9.9.* *Show the cardinality of the Cantor set is* $\mathfrak{c}$.

Now we consider the size or *measure* of the Cantor set. We consider the lengths of the interval that have been removed. The first iteration removes the middle third, which has length 1/3. The second iteration removes the middle thirds of the two remaining intervals, so another 2/9 is removed. The third iteration removes 4 intervals each of length 1/27. The sum of the lengths of all the interval removed by the whole process is

$$\sum_{n=1}^{\infty} \frac{2^{n-1}}{3^n}.$$

This is a geometric series with first term 1/3 and ratio 2/3, so

$$\sum_{n=1}^{\infty} \frac{2^{n-1}}{3^n} = \frac{1/3}{1-2/3} = 1.$$

This tells us that although the Cantor set is uncountable it has measure 0. This means it contains no intervals.

The Cantor set also has the property for every point x in the set and any distance ϵ, we can find another element of the Cantor set within distance ϵ of x. This is sometimes expressed as saying that every point in the Cantor set is a limit point. To see this let x have ternary expansion $0.a_1a_2a_3\ldots$. Choose a value of n such that $1/3^n < \epsilon$. We consider the first n digits of the ternary expansion of x, $0.a_1a_2a_3\ldots a_n$. Now consider $0.a_1a_2a_3\ldots a_n000\ldots$ and $0.a_1a_2a_3\ldots a_n222\ldots$. These both belong to the Cantor set and are distance less than ϵ from x. At least one of these is distinct from x.

9.2 FORMAL SYSTEMS

In 1900, the International Congress of Mathematicians invited David Hilbert to give a talk on what he thought were the most important mathematics problems for the new century. Hilbert was one of the world's leading mathematicians and a great supporter of Cantor. He thought that Cantor's work on infinity and the results on non-Euclidean geometry were some of the highlights of the nineteenth century. Hilbert listed twenty three unsolved problems that he considered the most important. The first problem concerned Cantor's work. He asked for a proof of the continuum hypothesis. His second problem concerned putting arithmetic on a sound axiomatic footing. There should be a list of axioms and rules of inference. Theorems should be proved by starting from the axioms and using the rules of inference at each stage.

Hilbert was worried about proofs using undefined terms or hidden assumptions being used. He wanted everything to be clearly defined.

FIGURE 9.3 David Hilbert

Formal Systems

A *formal system* consists of the following four things:

1. A list of symbols that will be needed—called an *alphabet*.
2. A *language* that tells us how to combine the symbols to form strings of symbols to form *well-formed formulas*, often called WFFs.
3. A set of axioms.
4. Rules of inference.

The axioms are all WFFs. The rules of inference tell us how to combine WFFs to form new WFFs. The last WFF in a chain of inference is called a *Theorem*, and the string of inferences is called a *proof*. In a formal system we ask whether a WFF is provable: Is it a theorem? We don't ask if a WFF is true or false.

There are certain properties that we would like our formal system to have: consistency, completeness and soundness.

Consistency

Given a set of axioms, it should be impossible to prove some statement and also the negation of the statement. A set of axioms for which it is impossible to derive contradictions is called *consistent*. In system that is not consistent, every statement is provable, so consistency is a basic requirement for any formal system that is going to be useful.

To prove a system is consistent it is enough to show that some statement is not provable.

Completeness

There are three possibilities for a statement:

- It can be proved.
- It can be disproved—there is a proof of the negation of the statement.
- It can neither be proved nor disproved.

If a system has the property that every statement can be either proved or disproved, then it is called *complete*. If there are statements that can neither be proved nor disproved, then it is *incomplete*.

Soundness

We will be describing a formal system for part of arithmetic. Given a statement about numbers, it is either true or false. We choose our initial axioms of the formal system to be true statements about numbers. We would like the rules of inference in the formal system to preserve truth, so that when we interpret a theorem derived in the formal system it gives a true statement about numbers.

We want the rules of inference to preserve truth. The theorems of the formal system should be true statements for any model (or interpretation) of the system. We define a formal system to be *sound* if the theorems are *universally valid*—true for every model of the system.

It is important to realize that consistency does not imply soundness. It is possible to construct a formal system in which the axioms are all true statements when interpreted as being about numbers, but some of the theorems that can be proved in the system give us false statements about numbers.

Since our models are consistent, no statement can be both true and false, our formal system must be consistent. So, soundness implies consistency, but is a stronger property.

First-Order Logic

Zeroth-order logic is the logic described by truth tables. It uses the connectives *and*, *or* and *not*. First-order logic includes zeroth-order logic and also allows the use of quantifiers, *there exists* and *for every*. So, for example, $2 < x < 3$ belongs to zeroth order logic, but the statement: *for any two natural numbers n and m, there exists a unique greatest common divisor*, belongs to first-order.

Much, but not all, mathematics can be formalized using first-order logic. Calculus and number theory can be formalized using it. Areas such as category theory and topos theory require higher order logics.

One nice thing about first-order logic is that it is sound. Anything we can prove using it will be universally valid.

The Hilbert Program

Initially, Hilbert's program for axiomatizing mathematics seemed promising.

In 1908, Ernst Zermelo gave an axiomatized version of set theory. It initially contained seven axioms. It was designed so that it was not possible to construct sets like the "set of all sets." It avoided the paradoxes. Zermelo also showed that it could be proved that any set could be well-ordered if you had a result now known as the *axiom of choice*.

The axiom of choice says that given any collection of sets, there is a way of choosing one element from each set in the collection. For a finite collection of sets, this is an obvious fact, but what if the number of sets in the collection was uncountable? There was some debate about whether the axiom of choice was needed and even if it was true for uncountably infinite collections.

In 1922, Abraham Fraenkel modified some of the axioms and added a new axiom on "replacement." This new set of axioms allowed all of Cantor's results, but still seemed to avoid contradictions. The Zermelo-Fraenkel axioms are usually known by the initials ZF, and ZFC if you include the axiom of choice. The ZFC axioms have stood the test of time and are now regarded as the accepted basic axioms for set theory.

Alfred North Whitehead and Bertrand Russell published the start of an enormous work to derive all of mathematics starting from logic. The first three volumes were published in 1910, 1912, and 1913. These volumes covered just set theory and numbers. There was to be a fourth volume on geometry, but the authors, discouraged about how long the process was taking and how much more mathematics needed to be covered, abandoned the project.

In the 1920s there were two major approaches for putting mathematics on a sound axiomatic footing. There was ZFC for set theory and Whitehead and Russell's "Principia Mathematica" for building all of mathematics starting from logic. Could it be proved that the sets of axioms underlying these were both complete and consistent?

ZFC and arithmetic

ZFC is a formal system for set theory that uses first-order logic. The arithmetic for natural numbers can be built within this framework. The simplest approach is the one given by John von Neumann. He began by defining 0 to be the empty set $\phi = \{\}$ and built the other numbers from there in the following way:

$$\begin{array}{rcl} 0 & = & \{\} \\ 1 & = & \{0\} \\ 2 & = & \{0,1\} \\ 3 & = & \{0,1,2\} \\ \vdots & \vdots & \vdots \\ n & = & \{0,1,2,\ldots,n-1\} \end{array}$$

We can then define $m < n$ by $m \in n$.

We won't show how to construct addition and multiplication, but ZFC with first-order logic has enough structure to enable both of these operations to be defined.

Kurt Gödel

Kurt Gödel proved some of the most stunning results in the history of mathematical logic. He would show that giving a complete and consistent set of axioms for set theory or even for arithmetic was not possible.

FIGURE 9.4 Kurt Gödel

A formal system has rules for constructing WFFs and rules of inference. We are interested in which statements can be proved from the axioms. We don't talk of truth, but of provability. Russell, in an essay from 1901 and later published as part of the book *Mysticism and Logic and Other Essays* writes:

> We start, in pure mathematics, from certain rules of inference, by which we can infer that *if* one proposition is true, then so is some other proposition. These rules of inference constitute the major part of the principles of formal logic. We then take any hypothesis that seems amusing, and deduce its consequences. *If* our hypothesis is about *anything*, and not about one or more particular things, then our deductions constitute mathematics. Thus mathematics may be defined as the subject in which we never know what we are talking about, nor whether what we are saying is true.

Earlier we said that if the underlying system of logic was first-order logic, then the provable statements are universally valid (true in all interpretations of the formal system). Gödel, in 1930, as part of his doctoral dissertation proved the converse: that if a statement is universally valid, then it is provable in the formal system. This result is called, somewhat confusingly, *Gödel's completeness theorem*. The completeness theorem is sometimes stated as showing the connection between syntactic provability and semantic truth. Russell's quote is about syntactic provability, Gödel shows that we can talk about truth and meaning, at least for first-order logic.

In 1931, Gödel published his two *incompleteness theorems*. The *first incompleteness theorem* states that no consistent system of axioms in which a certain amount of arithmetic can be carried out is complete.[2] There will always be statements that cannot be proved or disproved. The *second incompleteness theorem* says that it is not possible to prove the consistency of the set of axioms from within the system.

Gödel's incompleteness theorems stunned the mathematical world. It meant Hilbert's goal of finding complete axiomatic systems of mathematics was not feasible.

In 1938, Gödel showed that both the axiom of choice and the continuum hypothesis are consistent with the other axioms of ZF. This means that if ZF is consistent, then it is not possible to disprove either the axiom of choice or the continuum hypothesis. It still left open the question of whether you could prove either of them from ZF. This question was answered in 1963 by Paul Cohen. He showed that neither of them could be proved from ZF. This meant that the continuum hypothesis and the axiom of choice are independent of the axioms of ZF.

Most mathematicians accept and use the axiom of choice. It's widely used in analysis, but it does have some strange consequences. It allows that construction of *non-measurable sets*. These, as the name implies, cannot be measured—they do not have a length, or volume. This might not initially seem problematic, but the *Banach–Tarski paradox*[3] gives a way of decomposing a three-dimensional ball of radius 1 into five non-measurable sets. These sets can then be reassembled to form two three-dimensional balls of radius 1. The process of reassembly only involves translations and rotations. The process has doubled the volume. This initially seems impossible, but it is a logical consequence of ZFC. Our intuition is led astray by thinking of the non-measurable sets as being three-dimensional solid objects, but they are not. They are strange sets of points that cannot be assigned a volume.

---

[2]Gödel was not able to prove his first inconsistency theorem as it is usually stated. He needed to use a condition (called ω-consistency) that is more restrictive than consistency. In 1936, J. Barkley Rosser showed how the theorem could be proved using consistency).

[3]Despite the name, the Banach–Tarski paradox is not a paradox. It is called paradox because of the surprising non-intuitive conclusion.

There is no general agreement about the continuum hypothesis. Cantor believed it true; Cohen believed it false. Logicians and set-theorists are still divided. Is it even sensible to talk about its truth? Unlike, the axiom of choice, it is not needed or used for most mathematics.

Gödel's Completeness and Incompleteness Theorems

There is often confusion about these theorems, part of this is due to using the word *completeness* with two different meanings. These theorems do not contradict one another. An example should help to clarify the situation.

We know that $\mathbb{Q}$, $\mathbb{R}$, and $\mathbb{C}$ are all fields. Suppose we construct a formal system using the axioms for a field and first-order logic. The proposition *P: There exists an x in the field such that $x \times x = 2$* can be constructed in the formal system. Can P be proved?

The statement asserts the existence of the square root of 2. This is true in $\mathbb{R}$ and $\mathbb{C}$, but not in $\mathbb{Q}$. Since the statement is not true in every interpretation, it is not universally valid. Gödel's completeness theorem tells us P cannot be proved in the formal system. Similarly, we see the negation of P is not universally valid and so cannot be proved.

The statement P cannot be either proved or disproved in the formal system, so the system is incomplete. If we wish, we can add either P or *notP* as a new axiom.

Exercise 9.10⋆. *Show that adding the axiom* P *results in a system that is not complete.*

Exercise 9.11. *Show that adding the axiom* notP *results in a system that is not complete.*

The idea behind Gödel's proof of the first incompleteness theorem is often presented with the assumption that the system is sound. (We will talk more about this later.) First, construct a statement G_F within the formal theory F that states G_F is not provable in F. Once this self-referential statement has been constructed, if there was a proof of it, you would have a contradiction and F would be inconsistent. Consequently there cannot be a proof of G_F from within the system. So G_F is true.

Can we prove $notG_F$, the negation of G_F? We have shown G_f is true, so $notG_F$ is false. Since the system is sound, we cannot prove a statement that is false, and so you cannot prove $notG_F$. The statement G_F is independent of the other axioms.

We know G_F is true in the standard interpretation, but independent of the other axioms in F. If we add G_F as another axiom to the formal system, it will preserve consistency, the usual interpretation of number will be an interpretation of the new system, and G_F will be trivially provable. However, we now have a new formal system F' and we form the statement $G_{F'}$. This statement will not be provable in F', but will be a true statement in our usual interpretation. Though we can keep adding extra axioms, we will never obtain a complete system.

But things are more complicated.

Suppose we construct a formal system for arithmetic using first-order logic. The sentence G_F can be constructed within the system. We have just said that G_f is true, but the consistency theorem tells us it is not valid. It is a true statement with our usual numbers. Since G_F is not valid, there must be non-standard models of arithmetic in which G_F is false. These non-standard models are studied by mathematical logicians. However, most mathematicians are quite content to work solely with the standard ones.

Soundness and Consistency

Our sketch of the idea behind the proof of the first incompleteness theorem is semantic. We argued *we cannot prove a statement that is false.* We are using the soundness property. If we replace the hypothesis of consistency with soundness we have what is sometimes called the *semantic* version of the first incompleteness theorem.

Theorem 9.4. *(Gödel's first incompleteness theorem—semantic version) Any sound formal system in which a certain amount of arithmetic can be carried out is incomplete.*

The semantic version of Gödel's first incompleteness theorem stated is weaker than assuming just consistency, but it is the version of most use to mathematicians. It is also a version that can easily be proved once we introduce the idea of computability and the halting problem. We will sketch the proof later.

For most mathematicians, soundness is the useful property. We like to talk of statements being true or false. Why does Gödel use consistency rather than soundness in the hypotheses for his theorems?

Soundness is a semantic property. It enables us to assign meaning to statements. However, it involves interpretations of the formal system. These are not part of the formal system, but outside it. Soundness cannot be defined within the system. Consistency is a syntactic property. It can be defined within the formal system. Consistency is also a weaker property than soundness, and most mathematicians want to prove the most general theorem possible.

Theorem 9.5. *(Gödel's first incompleteness theorem) Any consistent formal system in which a certain amount of arithmetic can be carried out is incomplete.*

Consistency and Truth

Consistency does not imply soundness. If we have a consistent system in which our axioms are true statements we cannot assume the theorems are true statements. However, there are certain theorems we can deduce that are true.

We will consider the *Goldbach conjecture.* This dates back to 1742, when Christian Goldbach conjectured that every even number greater than 2 is equal to the sum of two prime numbers. This remains unsolved.

If the Goldbach conjecture is false, then there is an even number that cannot be written as the sum of two primes. If this number exists, we can find it by checking each even number. We can write a computer program that takes n, finds all the primes less than n, and tests whether n can be written as the sum of two of them. If the Goldbach conjecture is false, the program will eventually find the smallest number that cannot be expressed as the sum of two primes.

We are not arguing that this is feasible, or talking about how long the process will take, but making the observation that if the Goldbach conjecture is false, then we can prove this by using a finite process.

The hypotheses of Gödel's first incompleteness theorem are that the system must be consistent and contain a certain amount of arithmetic. The *certain amount of arithmetic* needed is exactly the amount needed to perform the search required to perform the disproof of Goldbach's conjecture if false.

Now suppose we have such a system and let *GC* be the statement in the system that corresponds to Goldbach's conjecture. We know that if *GC* is false, there will be a proof in the system. Now suppose that we can prove *GF*, then we can deduce it must be true, because if it was false there would be a proof, and so we could prove both *GC* and *notGC* but this is not possible because the system is consistent.

Exercise 9.12* *The Goldbach conjecture is unsolved. It is possible that it can be neither proved nor disproved. Could there be a proof of this?*

Gödel's proof uses these ideas. He constructs an arithmetical statement G_F that can be interpreted as *G_F is not provable in the formal system F*. If G_F is false then it can proved false by a finite search within F. As before, we can deduce that G_F is true and there is no proof of G_F. We know $notG_F$ is false. Is it possible to have a proof of *notGF*? We showed this is not possible if the system is sound, but we are only allowed to assume consistency and consistency does not rule this out.

Gödel decided to use a condition that is stronger than consistency but weaker than soundness—ω-consistency. We know that *notGF* is false, but that if it was true, there would be a finite search proof. Gödel showed statements like this cannot be proved assuming ω-consistency.

Rosser was able to weaken the hypothesis from ω-consistency to consistency by constructing a more complicated sentence R_F. This also has the property, that if false, it can be proved false by a finite search within F. The sentence is designed so that you can show $notR_F$ is false and also not provable using only consistency.

The Entscheidungsproblem

In 1928, Hilbert and Wilhelm Ackerman asked for an algorithm[4] that, given an axiomatic system, would take a statement as input and tell you whether or not it was universally valid. This is known as Hilbert's Entscheidungsproblem (decision problem). The decision problem can be thought of as an algorithm where you input any statement in the theory and the algorithm will tell you whether or not it can be proved from the axioms.

After Gödel proved his incompleteness theorems, it was clear that given an axiomatic system, there were two classes of statements: those that could be either proved or disproved from the axioms, and those that could neither be proved or disproved. If Hilbert's decision algorithm existed, given a statement, you could use the decision algorithm to tell which class of it belonged to. If it was provable, or its negation was provable, you could start work on proving it. If neither the statement or its negation were not provable, you knew it was independent of the other axioms. You could then decide whether or not to include it as a new axiom, but you would not waste time in trying to find a proof. Hilbert's algorithm, if it existed, would be an extremely useful tool.

Hilbert was asking for the construction of the algorithm. When he posed the problem, he believed such an algorithm existed. After Gödel's incompleteness theorems, other mathematicians were not so sure that it did. Both Alan Turing and Alonzo Church set out to show it was not possible to construct an algorithm for the decision problem. They worked independently, unaware of the other's work. Church was first, giving a talk at a meeting in 1935 and then publishing a paper in April 1936. Turing rushed his proof to print in the latter part of 1936.

Turing's proof was accepted for publication because of its originality. The ideas in his paper would later become the foundation for theoretical computer science.

9.3 COMPUTATION

Algorithms have always been part of mathematics. There are step-by-step methods for the calculation of greatest common divisors, of finding derivatives, and so on. But there was no definition of *algorithm*. To prove Hilbert wrong by showing no algorithm existed, both Turing and Church needed to start

[4]The word *algorithm* was not widely used at the time. *Effective procedure* was the term usually used.

by giving a definition of the term. Turing did this by defining a theoretical machine that is now known as a *Turing machine.*

Turing Machines

Turing described what he felt was the simplest mechanism that can perform any step-by-step calculation performed by humans. When we do a calculation by hand, it is often done with paper and pencil. Turing argued that the two-dimensionality of paper was not necessary, and that we could work with a one-dimensional tape. To do calculations with a tape, we need to be able to read and write on the tape.

He listed the basic components:

1. There is a finite *alphabet* consisting of all the possible symbols that can be written on the tape.
2. There is an infinite one-dimensional tape divided into cells. Each cell is either blank or can contain one letter from the alphabet.
3. Only a finite number of cells are not blank.
4. There is a tape head that at each step can move one cell to the left or to the right.
5. The tape head can read the current cell and can overwrite the symbol with another from the alphabet.

Before we proceed, we should explain the necessity of the infinite tape. When we perform a calculation using paper and pencil, we start with a blank page. If we fill the page with symbols, we start a new page. Turing didn't want the complication of adding new bits of tape, so he decided to have an infinite tape, but it can only have finitely many symbols written on it. All the other cells are blank.

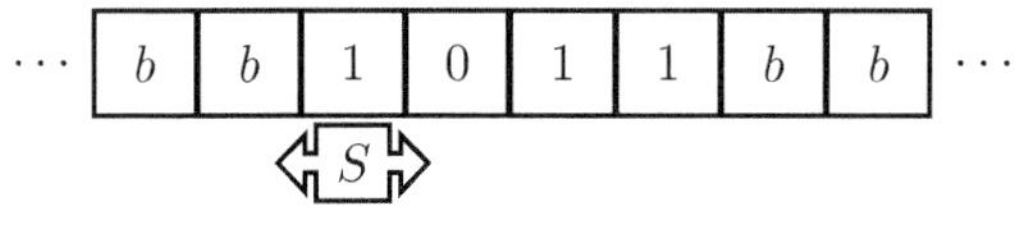

FIGURE 9.5 Turing machine tape and tape head

Figure 9.5 depicts the tape with the 1011 written on it. The b denotes blank cell. The tape head is reading the leftmost 1.

The machine can be in a finite number of *states.* We indicate the state of the machine by writing its name on the tape head. Figure 9.5 shows the machine is in state S.

These machines are programmed by listing a set of rules that tell it exactly what to do given the symbol it is reading and state the machine is in. For example, in figure 9.5 the machine is in state S and reading a 1. There will be a rule specifying what should be written over the 1, whether the tape head should move one step to the left or right, and what is the new state of the machine. We can write this down succinctly as a quintuple. For example, we might have a rule that says: If the machine is in state S and reading a 1, replace the 1 with a 0, move the head one step to the right and enter state B. We can write this rule succinctly as the quintuple $(S, 1, 0, R, B)$, where we use L and R in the fourth entry to denote 'left' and 'right,' respectively.

We also need to know which state the machine starts in and where the tape head is at the beginning. We will denote the starting state by S and assume the tape head is at the leftmost symbol that is not blank. Figure 9.5 shows the machine at the start of a computation.

Finally, certain states are denoted as final states. Once a machine enters one of these states it halts, ending the computation. An example, should help clarify how these machines work.

Turing machines are not practical computing devices, they were designed to be as simple as possible. Turing was breaking computation down into its elemental components.

FIGURE 9.6 Alan Turing

Turing Machine to Add 1 to a Binary Number

We will construct a Turing machine that takes as input a binary number and ends the computation with the number plus one on the tape. Referring back to figure 9.5, we can think of 1011 as the starting binary number. When we add 1 to this we do the following:

1. We look for the rightmost non-blank symbol. It is a 1.
2. We replace this 1 with a 0 and we look at the next symbol to the left. (We are carrying a 1.)
3. We replace this 1 with a 0 and we look at the next symbol to the left. (We are carrying a 1.)
4. We replace the 0 with a 1, and halt the computation with 1100 written on the tape.

Initially we move the tape head to right in order to find the rightmost symbol. Then there is the adding process. If the tape head is reading a 0, it should replace it with a 1 and then halt; it has finished the computation. If the tape head is reading a 1, it should replace it with a 0 and move one step to the left. There is one final case to consider. If during the adding process, the tape head reads a blank, it should replace it with a 1 and then halt.

We let S denote the starting state. It's the state that looks for the rightmost symbol, and let A denote the adding state. Finally, will let H denote the state that ends the computation.

In the starting state, there are three possibilities for what the tape head could be reading, 0, 1, or b. We list the quintuples:

$$(S, 0, 0, R, S), (S, 1, 1, R, S), (S, b, b, L, A)$$

The first two leave the binary string as it is and keeps the tape head moving the right. The third quintuple describes what happens when the tape head has gone past the last non-blank symbol on the tape and is now reading the first blank symbol to the right of the binary string. The tape head should now start to move to the left and enter the adding state.

The quintuples for state A are:

$$(A, 0, 1, L, H), (A, 1, 0, L, A), (A, b, 1, L, H)$$

The first and third of these quintuples tell the machine to halt, so there is no direction the head should move. I chose L, but R is also okay. Figure 9.7 shows the machine adding 1 to the binary number 1011 after starting as shown in figure 9.5.

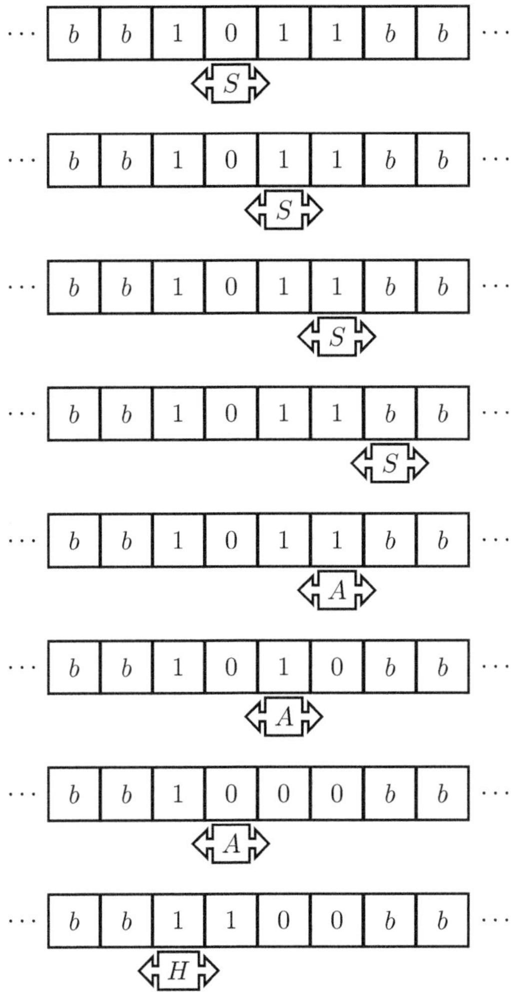

FIGURE 9.7 Adding 1 to 1011

Exercise 9.13. *A string of parentheses is balanced if each opening parenthesis is partnered with a closing parenthesis and the pairs are nested properly. For example (()()) is balanced, while (()))(is not. Design a Turing machine that takes inputs a string of parentheses and determines whether or not it is balanced.*

Church–Turing Thesis

Both Turing and Church needed to give a definition of *algorithm*. Turing claimed that Turing machines could perform any algorithm. Church defined algorithms in terms of a formal system of mathematical logic for functions, called the lambda calculus. (Functional programming is based on it.)

After Church published his proof, Turing realized that he needed to compare his definition to Church's. He showed that both definitions were equivalent. Two different approaches that yielded equivalent definitions helped to support the feeling that the definitions were correct. Since this time, various definitions have been given, and there is universal agreement that both Church and Turing have defined algorithms correctly.

The Universal Turing Machine

Turing machines are defined as lists of quintuples. The machine for adding 1 to a binary number is defined by:

$$(S,0,0,R,S),(S,1,1,R,S),(S,b,b,L,A),(A,0,1,L,H),(A,1,0,L,A),\\ (A,b,1,L,H).$$

If someone were to send us the quintuples along with the binary number 1011, we could perform the computation.

There is nothing special about this example. If we are sent any set of quintuples describing a Turing machine and an input string, we can perform the appropriate computation. If we want to, we can draw the tape and the tape head, but this is not needed. There is an *algorithm* that tells us what to do at each stage.

Once we are convinced that there is an algorithm that takes the quintuples describing a Turing machine and an input and then runs the machine on the input, we can apply the Church–Turing thesis. There must be a Turing machine that takes quintuples describing Turing machines and their inputs and emulates the Turing machine described quintuples on the input. We are led to the striking observation that *there must be a Turing machine that can emulate any other Turing machine on any input.* Such a machine is called a *Universal Turing Machine.*

This approach using the Church–Turing thesis was not the approach taken by Turing. He showed how to construct a Universal Turing Machine. He listed the quintuples. However, giving a full description of a Universal Turing Machine is complicated. We will content ourselves with the knowledge that the Church–Turing thesis tells us that they exist.

The Universal Turing machine can be compared to a modern computer. The inputted Turing machine corresponds to inputting a program and its associated input to data, both the program and data are treated as strings of symbols that are read into storage. This observation explains why Turing's work forms the basis of theoretical computer science.

The Halting Problem

Our example of a Turing machine for adding 1 to a binary number halts in every case, but not all Turing machines will halt on every input. For example, the Turing machine that has only one state and moves the tape head to the right,

$$(S, b, b, R, S),$$

never halts when started on a blank tape. More complicated Turing machines will halt on some inputs and not halt on others.

Can we design a Turing machine, that takes as input any Turing machine T along with its input I and tells us whether or not T halts on input I? This is the *halting problem.*

Suppose that this is possible. We will call this machine M.

$$M(T, I) \text{ outputs } YES, \text{ if machine } T \text{ halts on input } I$$

$$M(T, I) \text{ outputs } NO, \text{ if machine } T \text{ does not halt on input } I$$

Turing pointed out that everything can be written in binary. The description of a Turing machine can be written as a string of binary digits and the input can be written as a string of digits. One of the key ideas underlying the Universal Turing Machine was to treat both the program and data as strings of binary digits. But this means that you can take a Turing machine T and run it taking its input the binary string for T. (This is the self-referential idea that Cantor used in Theorem 9.3 and Gödel used in his first incompleteness theorem.)

We can slightly modify our machine M to give machine H:

$$H(T) \text{ outputs } YES, \text{ if machine } T \text{ halts on input } T$$

$$H(T) \text{ outputs } NO, \text{ if machine } T \text{ does not halt on input } T$$

We can slightly modify this machine so that instead of halting and printing YES, it goes into a loop and never halts.

$$N(T) \text{ does not halt, if machine } T \text{ halts on input } T$$

$$N(T) \text{ halts and outputs } NO, \text{ if machine } T \text{ does not halt on input } T$$

What happens if we run machine N on input N?

$$N(N) \text{ does not halt if machine } N \text{ halts on input } N$$

$$N(N) \text{ halts if machine } N \text{ does not halt on input } N$$

We obtain the nonsensical result that it does not halt if it halts and that it halts if it does not halt.

We conclude that our initial assumption that the machine M telling us whether or not a Turing machine T halts on I must be incorrect. There does not exist an algorithm that will tell us whether Turing machines will halt on their input. Of course, given a specific machine and a specific input, there might well be a way of determining whether it halts. The halting problem tells us there is no algorithm that works for all machines on all inputs.

The Halting Problem and Proving Gödel's First Incompleteness Theorem

In the previous sections, we showed how to assign strings of 0s and 1s to both Turing machines and their inputs. We can think of these as positive integers. (Perhaps inserting an additional 1 at the start of these strings and then thinking of the string being a positive integer written in binary.) The statement "Turing machine M halts on input I" becomes a statement about two positive integers.

We restate the semantic version of Gödel's first incompleteness theorem and sketch the proof.

Theorem. *(Gödel's first incompleteness theorem—semantic version) Any sound formal system in which a certain amount of arithmetic can be carried out is incomplete.*

Proof. Suppose, for a contradiction, that the system is complete. Then the question of whether M halts on I can either be proved or disproved. There is a proof in either case.

A proof in a formal system starts with the axioms and ends after a finite number of applications of the rules of inference. Any such proof can be found by a systematic search. We can construct an algorithm to do this.

Now we have an algorithm that solves the halting problem, but we know this is not possible. Our initial assumption that the system is complete must be false.

□

9.4 P AND NP

Our theory of computation has considered questions of what is theoretically possible to compute and what is impossible to compute. We have not mentioned feasibility. For example, we said that if Goldbach's conjecture is false we can prove this by running a program to search for counterexamples. If the conjecture is false, the program will halt after a finite amount of time. But a finite amount of time could be a very long time.

Time-complexity is a measurement of how long it takes for an algorithm to finish its computation. Clearly, some computers are faster than others, so a sensible measurement is the number of elementary operations that have to be performed during the computation. Consider sorting a list of numbers according to size. The number of operations will depend on the length of the list. The time required will be a function of the size of the input. We want a way to classify these functions.

Big O Notation

We want a way of describing how the length of time increases as the size of the input increases. We say that a function $f(n)$ has *order* $g(n)$, written $f(n) = O(g(n))$, if we can find a numbers M and N such that $f(n) \leq Mg(n)$ whenever $n \geq N$. For example, if the time function is polynomial in n, the highest power of a polynomial dominates the others. For $f(n) = 3n^2 + 2n + 1$, we can write $f(n) = O(3n^2)$, or even more succinctly $f(n) = O(n^2)$. If we use bubble sort for our list of n items, we can show its time complexity is $O(n^2)$.

The Class P

We say that an algorithm takes polynomial time if it is bounded by a polynomial. This can be stated as:

$$f(n) \text{ is polynomial time if there is a } k \text{ such that } f(n) = O(n^k).$$

Addition, subtraction, multiplication, division and the Euclidean algorithm are all polynomial time. If there is an algorithm for solving a problem in polynomial time the problem is often called "easy." These algorithms usually work well in practice. We denote the class of problems that can be solved in polynomial time by P.

We need to be a careful about the meaning of the 'size' of the input. If we consider Turing machines, the size of the input is the number of cells that are not blank at the start. If we take the symbols as either 0 or 1, then we could think of the input as being a number written in binary, but note that the size is not this binary number, but the number of digits that the binary number has. When we say that addition, multiplication, division are polynomial time, we are talking about how time increases when the number of digits increases.

An example of a problem that is not known to belong to P is factorization. The best algorithms are faster than 2^n but slower than any polynomial in n. Again, we need to be careful here; n is the number of digits in the number we are factoring. We are not factoring n.

The Class NP

The initials NP stand for nondeterministic polynomial time[5]. These are problems which you can *verify* in polynomial time. For example, we don't know how to do factorization in polynomial time. However, if someone tells you they have the answer, you can verify by multiplying the primes to make sure you get the correct original number. Factorization belongs to class NP.

Given a graph, finding a Hamiltonian cycle—the problem of finding a path that visits each vertex exactly once—belongs to NP. If someone hands you a solution, you can quickly verify it is correct. It can be verified in polynomial time, but nobody has found a polynomial-time algorithm for finding Hamiltonian cycles.

Many problems that belong to NP seem to be intractable.

Does P Equal NP?

If you have an algorithm that finds solutions in polynomial time, then you can verify it in polynomial time. So, all problems that belong to P also belong to NP. But what about the converse? Do all problems that belong to NP also belong to P? Are P and NP two names for exactly the same collection of problems?

Most people believe this is not the case. They believe there are problems in NP that are not in P. Initially it seems that proving NP is not equal to P should be easy. But it isn't. Nobody has found a proof that NP and P are not equal. Nobody has found a proof that they are. It remains the most important unsolved problem in computer science.

Proving that P equals NP would revolutionize our understanding of algorithms. It would tell us there are polynomial-time algorithms for solving problems that are currently considered intractable. Many of these NP problems are of practical use, knowing there are polynomial-time algorithms would have a major impact.

In 2000, the Clay Mathematics Institute, in the tradition of Hilbert's problems, listed seven problems they considered most important unsolved problems for the new millennium. These problems are called the *Millennium Problems*. Each problem comes with a million dollar award for the first correct solution. One of these is the question of whether P equals NP.

Within NP, there are a collection of problems known as *NP-complete*. If you can find a polynomial-time algorithm for solving any NP-complete problem, you not only show that that problem belongs to P, but every problem, in NP belongs to P. The Hamiltonian cycle problem is NP-complete. If you could

[5]The word *nondeterministic* comes from nondeterministic Turing machines. These machines are allowed to enter multiple states at each step.

find a polynomial-time algorithm for solving it, you would also prove that P equals NP.

Public Key Encryption

Internet traffic is encrypted. When communicating online, it is important to have the messages encrypted. The standard methods of encryption use *symmetric keys*. Both the sender and the receiver use the same key. Typically the key is a string of 128 or 256 binary digits. Given the key, there is a convoluted way of encrypting messages. This involves using the initial key to generate other keys, using bitwise addition to add the keys to chunks of the message, and performing various substitutions and permutations. Though there are several steps in the encryption process it is fast. It is also fast to decrypt. If you have the key, the encryption process can be reversed, quickly giving the original message. This method of encryption is believed to be secure.

Symmetric key encryption needs both parties to have the same key. This raises the question of *key distribution*. At the start of the communication, how can one party send the key securely to the other party? One way of doing this is to use RSA encryption.

RSA comes from the first initials of the last names of Ron Rivest, Adi Shamir and Leonard Adleman, who described the algorithm in 1977. The method involves three numbers e, d and n. The numbers e and n are public. The number d is kept secret.

Suppose that I want to send you a key that we will use as symmetric key for communication between us. This key is a binary string of 128 bits, but it can be thought of as a number written in binary. We will denote it by K. To encrypt the key, I find your public encryption key e and modulus n, then calculate $K^e \mod (n)$ and send it to you. You take the number you sent and calculate $(K^e)^d \mod (n)$. The result is the original number K. We now describe how the numbers are chosen, why it works and why it is secure.

The process begins by your computer finding two large primes p and q and then multiplying them to give the number n. The security of this method comes from the fact that finding large primes and multiplying them together are quick processes, but factorizing n without knowing the primes is hard. We already commented that multiplication can be done in polynomial time and there is no known polynomial-time algorithm for factorization. It might seem surprising that finding large primes is easy, but there are ways of randomly generating large numbers and using primality tests that do not use factorization.

Your computer then finds Euler phi function $\phi(n) = (p-1)(q-1)$. Next, it finds a number e that is coprime to $(p-1)(q-1)$. It can use the Euclidean algorithm to check this and in so doing find a number d such that

$$ed \equiv 1 \mod (p-1)(q-1).$$

Another way of expressing this congruence is to say there is an integer k such that

$$ed = 1 + k(p-1)(q-1) = 1 + k\phi(n).$$

The primes chosen for RSA are much larger than our key K. This tells us that K is coprime to n. Euler's theorem says that $K^{\phi(n)} \equiv 1 \mod (n)$, which tells us

$$K^{ed} \equiv K^{1+k\phi(n)} \equiv K \mod (n).$$

This is the important identity.

To summarize, the numbers e and n are public. You keep d secret. To encrypt K, I calculate $K^e \mod (n)$ and send you this number. You then raise this to the dth power modulo n and it gives you back K. We now both have a key to use as a symmetric key.

RSA is widely used. However, there is a search for a replacement. The security of RSA depends on the number n being hard to factor. In 1994, Peter Shor found a quantum algorithm that can factor large numbers in polynomial time on a quantum computer. If large enough quantum computers could be built, RSA could be broken.

Shor's algorithm initiated an area known as *post-quantum cryptography*[6] looking for algorithms that can resist attacks by quantum computers. The National Institute of Science and Technology in the United States has started a competition for a replacement for RSA that cannot be broken by quantum attacks.

Suggestion for Further Reading

Chris Bernhardt. *Turing's Vision: The Birth of Computer Science.* MIT Press, 2016.

Paul Halmos. *Naive Set Theory.* Dover Publications; Reprint Edition. 2017

Torkel Franzén. *Gödel's Theorem.* A. K. Peters. First Edition 2005

[6]This should not be confused with *quantum cryptography*—cryptography using quantum computers. Quantum key exchange methods using quantum computers are highly secure.

Solutions to Starred Exercises

1.1 This follows from the fact that $(x_1 - x_0)^2 = (x_0 - x_1)^2$ and $(y_1 - y_0)^2 = (y_0 - y_1)^2$.

1.4 The diagram with the radius drawn is below. The center of the semicircle is denoted by 0. The triangle OAC is isosceles, so the base angles, denoted by α are equal. Similarly, the triangle OBC is isosceles with equal base angles (denoted by β). The sum of the angles of the triangle ABC add to $2\alpha + 2\beta$. Since the sum of the angles of a triangle is $180°$, we obtain $\alpha + \beta = 90°$.

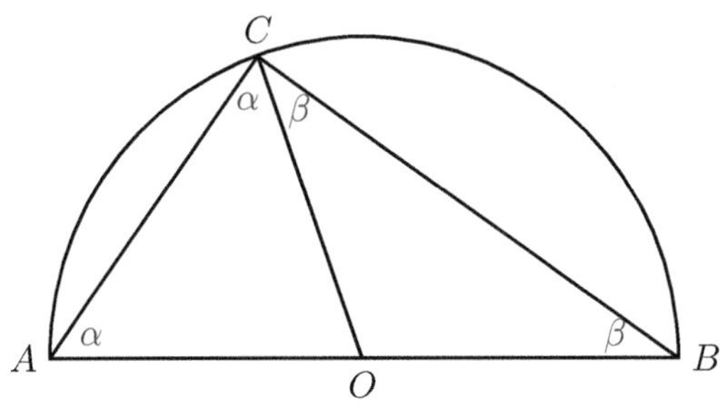

1.5 The quadratic formula tells us the two roots of $x^2 - ax + b^2 = 0$ are $x = a/2 \pm \sqrt{a^2/4 - b^2}$. The Pythagorean theorem shows the length of PD is $\sqrt{a^2/4 - b^2}$. We know $|AP| = |PB| = a/2$. So, $|AD| = a/2 + \sqrt{a^2/4 - b^2}$ and $|DB| = a/2 - \sqrt{a^2/4 - b^2}$.

1.9 Triangles ABC and DAC are both isosceles. They have equal base angles and so are similar. By similarity, $|BA|/|AC| = |AC|/|DC|$.

1.12 The equation of the semicircle is

$$(x - c/2)^2 + y^2 = (c/2)^2$$

or

$$x^2 - cx + y^2 = 0.$$

Substituting $y = x^2/b$ and simplifying shows that the non-zero root x satisfies

$$x^3 + b^2x = b^2c.$$

1.13 Let $x_2 > x_1 > 0$. Then

$$(x_2^3 + b^2x_2) - (x_1^3 + b^2x_1) = x_2^3 - x_1^3 + b^2(x_2 - x_1)$$
$$= (x_2 - x_1)(x_1^2 + x_1x_2 + x_2^2) + b^2(x_2 - x_1) > 0.$$

That is to say, the expression $x^3 + b^2x$ is strictly increasing in x and hence can take on any given value at most once.

2.1 Let d denote the gcd of a and b. The Euclidean algorithm shows that d belongs to

$$S = \{ax + by \,|\, x, y \text{ are integers}\}.$$

Let c denote the smallest positive integer in S. Since d divides both a and b, it must divide every number in S. In particular, d must divide c. We know that $c \leq d$ and d divides c, so c must equal d.

2.4 If p divides $(ab)c$ and does not divide c, then Euclid's Lemma tells us it must divide ab. Applying the lemma once more tells us p must divide either a or b.

2.6 Let p be a prime dividing mn. Since m and n are coprime, p must divide either m or n, but not both. We look at the case when p divides m. In this case p divides m^2 but not n^2, so it cannot divide $m^2 \pm n^2$. A similar argument works when p divides n but not m.

2.8 Let $d_1 = 1 < d_2 < \cdots < d_m = p$ denote the positive divisors of p and $e_1 = 1 < e_2 < \cdots < e_n = q$ the positive divisors of q. Since p and q are coprime, any positive divisor of pq can be written as a positive divisor of p multiplied by a positive divisor of q in exactly one way.

$$\sigma(pq) = \sum_{i,j} d_i e_j = \sum_i d_i \sum_j e_e = \sigma(p)\sigma(q).$$

2.10 If p is not prime, we can write is as $p = ab$, where a and b are positive integers greater than 1. Using the fact that

$$x^b - 1 = (x-1)(x^{b-1} + x^{b-2} + \cdots x + 1)$$

on the expression $2^p - 1 = (2^a)^b$ gives the factorization

$$2^p - 1 = (2^a - 1)\left((2^a)^{b-1} + (2^a)^{b-2} + \cdots (2^a) + 1\right)$$

showing that if p is not a prime then $2^p - 1$ is composite.

2.12 If $a \equiv b \mod (n)$, then $b-a = kn$ for some integer k. If $c \equiv d \mod (n)$, then $c-d = ln$ for some integer l. Now $(b+d)-(a+c) = (k+l)n$, so n divides $(b+d)-(a+c)$ telling us $a+c \equiv b+d \mod (n)$.

2.15 If $ja \equiv ka \mod (n)$, then n divides $(k-j)a$. Since n and a are coprime, n must divide $k-j$, so they belong to the same congruence class.

2.20 If n is composite, we can find integers r and s with $1 < r \leq s < n$ and $n = rs$. If $r \neq s$, they will both appear in the product $1 \times 2 \times 3 \times \cdots \times (n-1)$, so $n = rs$ will be a divisor. If $r = s$, then $n = r^2$. Since $n > 4$, $r \geq 3$. This implies $r < 2r < n$. Since both r and $2r$ appear in the product $1 \times 2 \times 3 \times \cdots \times (n-1)$, n must be a divisor.

4.1 Let $p(z) = a_n z^n + a_{n-1} z^{n-1} + \ldots a_n$ be a polynomial with real coefficients. If z is a complex root of p then

$$0 = \overline{p(z)} = a_n \bar{z}^n + a_{n-1} \bar{z}^{n-1} + \ldots a_0$$

hence the complex roots of p occur in complex conjugate pairs. Denote the real and complex roots $z_1, \ldots, z_n$ of p by $r_1, \ldots, r_s$ and $\alpha_1 \pm i\beta_1, \ldots, \alpha_t \pm i\beta_t$. Then

$$\begin{aligned} p(z) &= a_n(z-z_1)\ldots(z-z_n) \\ &= a_n(z-r_1)\ldots(z-r_s)(z-\alpha_1-i\beta_1)(z-\alpha_1+i\beta_1)\ldots \\ &\quad (z-\alpha_t-i\beta_t)(z-\alpha_t+i\beta_t) \\ &= a_n(z-r_1)\ldots(z-r_s)[(z-\alpha_1)^2+\beta_1^2]\ldots[(z-\alpha_t)^2+\beta_t^2]. \end{aligned}$$

4.4 The Tschirnhaus substitution $x = y - 1$ yields

$$y^3 + 4y + 2 = 0.$$

Applying Cardano's formula (4.14) with $p = 4$ and $q = 2$ gives (to 4 decimal places) $y = -0.4735$. Thus $x = -1.4735$.

4.7 Denote $z = 1 + i$. Then $|z| = \sqrt{1^2 + 1^2} = \sqrt{2}$ and $\arg(z) = \arctan 1 = \pi/4$. Thus the four fourth roots of z are

$$\begin{aligned} &\sqrt[8]{2}(\cos \pi/16 + i \sin \pi/16)) \\ &\sqrt[8]{2}(\cos(\pi/16 + \pi/2) + i \sin(\pi/16 + \pi/2)) = \sqrt[8]{2}(\cos 9\pi/16 + i \sin 9\pi/16) \\ &\sqrt[8]{2}(\cos \pi/16 + \pi) + i \sin(\pi/16 + \pi)) = \sqrt[8]{2}(\cos 17\pi/16) + i \sin 17\pi/16) \\ &\sqrt[8]{2}(\cos \pi/16 + 3\pi/2) + i \sin(\pi/16 + 3\pi/2)) = \sqrt[8]{2}(\cos 17\pi/16) \\ &\quad + i \sin 25\pi/16). \end{aligned}$$

4.9 We show one of the cases:

$$\begin{aligned}V_i(d,c,a,e,b) &= d+\omega^i c+\omega^{3i}a+\omega^{2i}e+\omega^{4i}b\\ &= \omega^{3i}(a+\omega^i b+\omega^{3i}c+\omega^{2i}d+\omega^{4i}e\\ &= \omega^{3i}V_i(a,b,c,d,e).\end{aligned}$$

3.2 Denote $S = F_1 + F_2x + F_3x^2 + \ldots$. Using the defining property (3.2), we have

$$S - xS - x^2S = F_1 + F_2x - xF_1 = 1.$$

3.3 Denote

$$\frac{1+\sqrt{5}}{2} = \phi_1, \quad \frac{1-\sqrt{5}}{2} = \phi_2.$$

Then

$$\frac{F_{n+1}}{F_n} = \phi_1\left[\frac{1-\left(\frac{\phi_2}{\phi_1}\right)^n}{1-\left(\frac{\phi_2}{\phi_1}\right)^{n-1}}\right].$$

Since $|\phi_2/\phi_1| < 1, \lim_{n\to\infty}(\phi_2/\phi_1)^n = \lim_{n\to\infty}(\phi_2/\phi_1)^{n-1} = 0$, and the result follows.

3.4

$$L_n = \left(\frac{1+\sqrt{5}}{2}\right)^{n-1} + \left(\frac{1-\sqrt{5}}{2}\right)^{n-1}.$$

3.5 $x = 1, y = 3, z = 4$.

3.6 Multiplying the first equation by c, the second equation by a and subtracting eliminates x from the equation and gives

$$(bc - ad)y = cp - aq. \tag{1}$$

If $bc \neq ad$, then this yields $y = (cp - aq)/(bc - ad)$ and substituting into either of the original equations and solving for x, we obtain a unique solution (x, y) to the system.

On the other hand, suppose $bc = ad$. If $cp \neq aq$ then equation (1) implies that the system is inconsistent hence has no solutions. Suppose $cp = aq$. We assume that in each of the equations at least one of the coefficients on the left-hand side is non-zero, otherwise the equation either results in inconsistency or vanishes. Assume without loss of generality that $a \neq 0$. Then the first equation has infinitely many solutions (x, y) where $x = (p - by)/a$. These solutions also satisfy the second equation since

$$cx + dy = c\left(\frac{p-by}{a} + dy\right) = \frac{cp - cby}{a} + cdy = \frac{aq - ady}{a} + dy = q.$$

5.1 Following the hint and using the algebraic identity $(a+b)(a-b) = a^2-b^2$, we have

$$\begin{aligned}\lim_{x\to 9}\frac{\sqrt{x}-3}{x-9} &= \lim_{x\to 9}\frac{\sqrt{x}-3}{x-9}\times\frac{\sqrt{x}+3}{\sqrt{x}+3}\\ &= \lim_{x\to 9}\frac{x-9}{(x-9)(\sqrt{x}+3)} = \frac{1}{6}.\end{aligned}$$

5.2 In order that f to be continuous at $x = 3$, it is first necessary that the left and right limits of $f(x)$ as x approaches 3, agree (so that $\lim_{x\to 3} f(x)$ exists). This implies

$$9 + 2c = 9 + c^2.$$

Thus $c = 0$ or 1. For these choices of c, $\lim_{x\to 3} f(x) = f(3)$, hence f is continuous at 3.

5.3 $y - 1/2 = \frac{7}{8}(x-1)$.

5.4 By inspection $f(2) = 21$, i.e., $f^{-1}(21) = 2$. By IFT, we have

$$(f^{-1})'(21) = \frac{1}{3x^2+4}\Big/_{x=2} = 1/16.$$

5.9 $e'(x) = \frac{1}{\log'(e^x)} = e^x$.

5.13 Assume $e = p/q$, so

$$\frac{p}{q} = 1 + \frac{1}{2!} + \frac{1}{3!} + \cdots + \frac{1}{q!} + \frac{1}{(q+1)!} + \cdots.$$

Following the hint, we have

$$p(q-1)! = q!\left(1 + \frac{1}{2!} + \frac{1}{3!} + \cdots + \frac{1}{q!}\right) + \frac{1}{(q+1)} + \frac{1}{(q+1)(q+2)} + \cdots.$$

It follows that the sum of the infinite series

$$S = \frac{1}{(q+1)} + \frac{1}{(q+1)(q+2)} + \frac{1}{(q+1)(q+2)(q+3)} + \cdots$$

is an integer. However

$$S < \frac{1}{(q+1)} + \frac{1}{(q+1)^2} + \frac{1}{(q+1)^3} + \cdots = \frac{1/(q+1)}{1-1/(q+1)} = \frac{1}{q}$$

and since there is no integer between 0 and $1/q$, this gives a contradiction.

5.14 The difference quotient at 0 for this function is $\frac{1}{h}e^{-1/h^2}$. Substituting $t = 1/h$, we have

$$\lim_{h\to 0}(1/h)e^{-1/h^2} = \lim_{t\to\infty} te^{-t^2} \le \lim_{t\to\infty}\frac{t}{1+t^2} = 0.$$

6.2 Using the Binomial Theorem and interchanging the order of summation, we have

$$\begin{aligned} e^{z+w} &= \sum_{k=0}^{\infty} \frac{(z+w)^k}{k!} \\ &= \sum_{k=0}^{\infty} \sum_{r=0}^{k} \frac{z^r w^{k-r}}{r!(k-r)!} \\ &= \sum_{r=0}^{\infty} \frac{1}{r!} \sum_{k=r}^{\infty} \frac{z^r w^{k-r}}{(k-r)!} \\ &= \sum_{r=0}^{\infty} \frac{z^r}{r!} \sum_{k=0}^{\infty} \frac{w^k}{k!} = e^z \cdot e^w. \end{aligned}$$

6.3 Let $z \in D$, so $|z| < 1$. Choose $r = (1 - |z|)/2$ and let D_z denote the disc

$$D_z = \{|w - z| < r\}.$$

Then by the triangle inequality, if $w \in D_z$ then

$$|w| \le |z| + |w - z| < |z| + r = (1 + |z|)/2 < 1.$$

Thus $D_z \subset D$.

6.5 Differentiating first in x and then in y, we obtain

$$\begin{aligned} UU_x + VV_x &= 0 \\ UU_y + VV_y &= 0. \end{aligned}$$

Together with the Cauchy–Riemann equations

$$\begin{aligned} U_x - V_y &= 0 \\ U_y + V_x &= 0 \end{aligned}$$

we have a homogeneous system of 4 linear equations in U_x, U_y, V_x, V_y. In matrix form,

$$\begin{bmatrix} U & 0 & V & 0 \\ 0 & U & 0 & V \\ 1 & 0 & 0 & -1 \\ 0 & 1 & 1 & 0 \end{bmatrix} \begin{bmatrix} U_x \\ U_y \\ V_x \\ V_y \end{bmatrix} = \begin{bmatrix} 0 \\ 0 \\ 0 \\ 0 \end{bmatrix}.$$

It follows from the fact that U and V cannot both vanish, that the coefficient matrix has non-zero determinant and hence is invertible. Thus the system has the unique solution $U_x = U_y = V_x = V_y = 0$, which implies that U and V are constant.

6.8 Suppose f is entire and $|f(z)| \leq M$, for all $z \in \mathbf{C}$. Then by (6.18) with $n = 1$ and Theorem 6.5, we have, for $R > |z|$,

$$|f'(z)| = \left|\frac{1}{2\pi i}\int_{C_R(0)} \frac{f(\xi)}{(\xi - z)^2}d\xi\right| \leq \frac{2\pi RM}{2\pi(R - |z|)^2} \to 0$$

as $R \to \infty$. Thus $f'(z) \equiv 0$, which implies f is constant.

6.10 Let $w_0 \neq 0$. By Picard's theorem, there exists z_0 such that $e^{z_0} = w_0$. Then $e^{1/z} = w_0$ for

$$z = \frac{1}{z_0 + 2n\pi i}, \quad n \in \mathbf{N}.$$

6.11

$$Res\left(\frac{z}{(z-1)(z-2)^2}, 1\right) = \frac{z}{(z-2)^2}\Big/_{z=1} = 1$$

$$Res\left(\frac{z}{(z-1)(z-2)^2}, 2\right) = \left[\frac{z}{z-1}\right]'(2) = -1.$$

Thus

$$\int_{|z=3|} \frac{zdz}{(z-1)(z-2)^2} = 0.$$

6.13 $\pi^2/8$.

7.1 Each edge connects two vertices. So each edge contributes 2 to the sum of the degrees of the vertices.

7.2 A graph is 1-regular if the degree of each vertex is 1. Each vertex is the endpoint of exactly one edge. This means each vertex is connected to exactly one other vertex, so the vertices can be paired. Each of the pairs corresponds to an edge, so the number of vertices is twice the number of edges.

7.4 We have to draw the graph in the plane with no edge crossings. One way of doing this is:

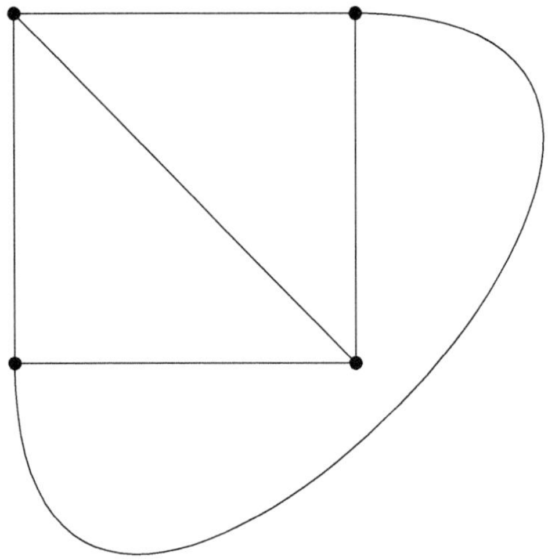

7.6 We have to draw the graph in the plane with no edge crossings. One way of doing this is:

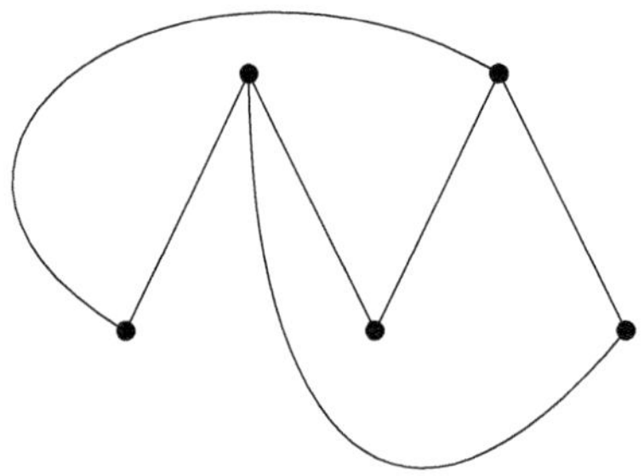

7.7 The vertex that has m vertices and no edges is an example.

7.8 For a connected tree there is a unique sequence of edges connecting any two vertices. Choose a vertex and give it a color. Every other vertex is connected to this colored vertex by either an even number of edges or an odd number. Use the same color for the vertices connected by an even number of edges and a different color for the vertices connected by an odd number of edges. This gives a coloring because vertices of the same color are separated by an even number of edges.

7.10 In a complete graph, any two vertices are connected by an edge, so no two vertices can have the same color. This means each vertex must have a distinct color.

8.2 Firstly, there are 10 different combinations of ranks that can be straights, depending on where the straight starts, ace - 10 (allowing ace to be both high and low). Then there are 4 possible choices of the suit for each rank, giving a total of $10 \cdot 4^5$ hands that are straights. Thus the probability of a straight is

$$\frac{10 \cdot 4^5}{\binom{52}{5}} = .0039.$$

8.4 6/7.

8.5 Write

$$X = \sum_{i=1}^{n} X_i$$

where $X_1, \ldots, X_n$ are independent Bernoulli random variables with parameter p. Then, by Theorem 8.1,

$$E[X] = \sum_{i=1}^{n} E[X_i] = np,$$

$$Var(X) = \sum_{i=1}^{n} Var(X_i) = npq.$$

8.6 By Theorem 8.1, $\bar{X}$ has mean 70 and standard deviation $5/\sqrt{100} = .5$. The sampled for $\bar{X}$ of 65, thus lies 10 standard deviations below the mean. According to Chebyshev's inequality, the likelihood of this happening is less than 1%. This provides strong evidence that the claim is correct.

8.7 Approximately 1/3000.

8.8 Note that $X_{10} = 4$ if there are 7 steps to the right and 3 steps to the right in 10 moves. The probability of this event is

$$\binom{10}{7}(0.4)^7(0.6)^3 = 04246.$$

8.9 Define

$$f(p) = pq = p(1-p),\ 0 \leq p \leq 1.$$

This function attains its maximum value when

$$f'(p) = 1 - 2p = 0,$$

i.e., $p = 1/2$. Since $f'(p)$ is positive in the region $(0, 1/2)$ and negative in $(1/2, 1)$ it follows that for $p \neq 1/2, f(p) < f(1/2) = 1/4$. (The result also follows from the vertex formula for a parabola.)

8.13 Assuming without loss of generality that the random walk starts at position 0, the position X_n at time n is given, following (8.10) by

$$X_n = \sum_{k=0}^{n} Z_k.$$

The means μ of the Z_k are $p - q > 0$. According to SLLN, $X_n/n \to \mu$ as $n \to \infty$, with probability 1. This implies that, with probability 1, there exists a time N such that for $n > N, X_n \neq 0$, i.e., X_n returns to 0 at most finitely many times. This implies that the random walk is transient.

9.2 The derivative of $g(x)$ is $g'(x) = -e^x/(1+e^x)^2$. This is always negative, so $g(x)$ is a decreasing function. It is straightforward to check

$$\lim_{x\to-\infty} g(x) = 1$$

and

$$\lim_{x\to\infty} g(x) = 0.$$

9.3 We will use binary expansions of real numbers, but as with decimals we have to be careful about numbers having two expansions. For decimal expansions, any number with an infinite tail of 9s can be rewritten to have an infinite tail of 0s. For binary expansions, any number with an infinite tail of 1s can be rewritten to have an infinite tail of 0s.

The cardinality of the half-open interval $[0, 1)$ is $\mathfrak{c}$. The set of infinite binary strings has cardinality $2^{\aleph_0}$. We construct one-to-one functions from each of these sets to the other and then use the Schröder–Bernstein theorem to conclude they have the same cardinality.

We can define a map f from infinite binary strings to the interval $[0, 1)$ by defining

$$f(b_1, b_2, b_3, \ldots, b_k \ldots) = \frac{b_1}{10} + \frac{b_2}{10^2} + \cdots + \frac{b_k}{10^k} + \cdots = \sum_{k=1}^{\infty} \frac{b_k}{10^k}.$$

This function is not onto, but it is one-to-one.

We can define a one-to-one function from $[0, 1)$ to binary strings by taking the binary expansion of the number that doesn't end in a string of 1s. Again, this is not onto, but is one-to-one.

9.4 A subset of S can be generated by going through the set of elements of S one by one and deciding whether to include it or not. There are two choices for each element. The set of all subsets can be generated by going through the $2^{|S|}$ choices.

9.9 Each point in the Cantor set can be written as an infinite string of 0s and 2s. We can map this to infinite binary strings by changing the 2s to 1s. This gives a bijection between the Cantor set and the set of infinite binary strings.

9.10 Both $\mathbb{R}$ and $\mathbb{C}$ satisfy the axioms for a field and also axiom P. The statement Q: *There exists an x in the field such that $x \times x = -1$* is true in $\mathbb{C}$ but not in $\mathbb{R}$, so the new system is not complete.

9.12 No. If the Goldbach conjecture is false, then as shown in the text it can be proven false. If we have a proof that it cannot be proven false, then it must be true.

Bibliography

[1] Scott Aaronson. *Quantum Computing Since Democritus.* Cambridge University Press, 1st edition, 2013.

[2] Chris Bernhardt. *Turing's Vision: The Birth of Computer Science.* MIT Press, 2016.

[3] Carl B. Boyer and Uta C. Merzbach. *A History of Mathematics.* Wiley, 3rd edition, 2011.

[4] David M. Burton. *The History of Mathematics: An Introduction.* McGraw Hill, 7th edition, 2010.

[5] Euclid. *Euclid's Elements.* Green Lion Press, 1st edition, 2002. Dana Densmore (editor), Thomas L. Heath (translator).

[6] Stephen D. Fisher. *Complex Variables.* Dover Publications, 2nd edition, 1999.

[7] Torkel Franzén. *Gödel's Theorem: An Incomplete Guide to its Use and Abuse.* A K Peters, 2005.

[8] Theodore W. Gamelin. *Complex Analysis.* Springer, 2021.

[9] Paul Halmos. *Naive Set Theory.* Dover Publications, reprint edition, 2017.

[10] G. H. Hardy. *A Mathematician's Apology.* Hawk Books, 1987.

[11] Nora Hartsfield. *Pearls in Graph Theory.* Dover Publications, 2013.

[12] Sir Thomas L. Heath. *A Manual of Greek Mathematics.* Dover Publications, 2003.

[13] James R. Kirkwood. *An Introduction to Analysis.* Chapman and Hall/CRC, 3rd edition, 2021.

[14] John W. Lamperti. *Probability: A Survey of the Mathematical Theory.* Dover Publications, 2003.

[15] Christopher Moore and Stephan Mertens. *The Nature of Computation.* Oxford University Press, 1st edition, 2011.

[16] Sheldon M. Ross. *Introduction to Probability Models.* Academic Press, 13th edition, 2023.

[17] Ian Stewart. *Galois Theory.* Wiley, 2nd edition, 1996.

[18] John Stillwell. *Mathematics and Its History.* Springer, 3rd edition, 2010.

[19] John Stillwell. *Elements of Mathematics: From Euclid to Gödel.* Princeton University Press, reprint edition, 2017.

[20] Dirk J. Struik. *A Concise History of Mathematics.* Dover Publications, 4th revised edition, 2011.

[21] Jean-Pierre Tignol. *Galois' Theory of Algebraic Equations.* World Scientific, 2nd edition, 2016.

[22] Richard J. Trudeau. *Introduction to Graph Theory.* Dover Publications, 2nd edition, 1994.

Index